SCIENTIFIC METHODOLOGY IN NINETEENTH CENTURY BRITAIN

Nineteenth-Century Science, Technology and Medicine: Sources and Documents

Meteorology in Nineteenth-Century Society
4 volume set
Edited by Aitor Anduaga

Nineteenth-Century British Microscopy
4 volume set
Edited by Meegan Kennedy

Nineteenth-Century Natural Theology
4 volume set
Edited by Adam R. Shapiro

The Science of Mind in the Nineteenth Century
4 volume set
Edited by Stephen T. Casper & Jon D. Goss

Science and Sound in Nineteenth-Century Britain
4 volume set
Edited by Edward J. Gillin

Scientific Methodology in Nineteenth Century Britain
4 volume set
Edited by Charles H. Pence

Women in Medicine in the Long Nineteenth Century
4 volume set
Edited by Claire Brock

Insects in Victorian Culture
4 volume set
Edited by John Clark

Electric Power in Victorian Britain
4 volume set
Edited by Nathan Kapoor

Women, Gender and Science in Victorian Culture
4 volume set
Leila McNeill & Anna Reser

SCIENTIFIC METHODOLOGY IN NINETEENTH CENTURY BRITAIN

Edited by
Charles H. Pence

Volume I
Building Philosophical Systems

Routledge
Taylor & Francis Group
LONDON AND NEW YORK

First published 2026
by Routledge
4 Park Square, Milton Park, Abingdon, Oxon OX14 4RN

and by Routledge
605 Third Avenue, New York, NY 10158

Routledge is an imprint of the Taylor & Francis Group, an informa business

British Library Cataloguing-in-Publication Data
A catalogue record for this book is available from the British Library

ISBN: 978-1-032-20488-8 (set)
ISBN: 978-1-032-20490-1 hbk (vol I)
ISBN: 978-1-003-26381-4 ebk (vol I)

DOI: 10.4324/9781003263814

Typeset in Times New Roman
by Apex CoVantage, LLC

CONTENTS

SERIES PREFACE

Nineteenth-Century Science, Technology and Medicine: Sources and Documents

GENERAL EDITORS' INTRODUCTION

This volume is part of the Routledge Historical Resources primary source collection, "Nineteenth-Century Science, Technology and Medicine: Sources and Documents." This series offers a fresh look at the historical development of science, technology, and medicine through the eyes of subject experts and specialists from around the world. The series provides curated access to significant primary-source documents on these subjects, collating and contextualizing these valuable resources so that they may ground not just scholarly research but undergraduate and graduate research and teaching as well. As well as being published in print and ebook formats, the contents of this and all other volumes in this series will be available through the Routledge Historical Resources database, where they will be presented in a fully integrated and searchable format, together with supplemental secondary sources and images and newly commissioned scholarly written and video essays.

Our editors have framed their volumes around themes they consider most pertinent to understanding nineteenth-century developments in their chosen field and its significance to broader changes in nineteenth-century British culture and society. Often the editors have chosen to provide short extracts to illustrate particular points or themes; in other cases, they have provided entire essays, book chapters, or periodical articles as appropriate. Each volume is prefaced by an introductory essay in which the editors review their topic and discuss the significance of their sources. They also frame each entry or collection of related entries with explanatory headnotes. Thus, each volume or set of volumes will provide a useful resource of easy-to-navigate, well-contextualized sources curated by an expert eye, which will serve as a useful introduction for those unfamiliar with the field. Since the editors build each volume to represent a unique, deeply considered scholarly perspective on the foundational primary sources for a particular subject, the series should also prove valuable to established scholars, even those with deep experience in the field. Indeed, given the range of topics covered and sources collated in this series, we anticipate that this project will foster an exciting and insightful interdisciplinary perspective on the histories and cultures of nineteenth-century

British science, technology, and medicine. We as general editors have learned a lot through our engagement with the editors who have contributed to this project, and we believe that wider audiences will, too.

Throughout the project we have asked editors to comment on the state of the existing historiographical debate, and to specify where their own work fits in. Some editors have chosen to do this through the provision of recommendations for further reading; others do so through explicit commentary on this topic in their introductory essays, and in the headnotes and footnotes to each of the sources they have selected.

The histories of science, technology, and medicine are wide and often disparate fields, even though they are often encountered by students in the context of a single university department. Although they have much in common, scholars of science, technology, and medicine also propose field-specific questions and approaches; these are represented in the project through the disparate interests of the range of scholars contributing to the series. We have also worked to bring the insights of literary scholars as well as historians to the project.

We have made every effort to cover major nineteenth-century themes and topics in the scientific and technological disciplines as well as to give space to specialist topics. Of course, as is inevitably the case in a project of this nature, which must rely on the availability of expert scholars in particular fields, there will remain gaps. We have attempted to address these where we can in the form of introductory essays and video essays in the database. We do believe, though, that the reader and researcher will find that this project presents the major themes, as well as many specialist topics, in detail, and with scholarly rigour.

We are very grateful to Routledge for their invitation to take on this project, and we would especially like to thank the commissioning editor Rachel Douglas, her successor, Laura Pilsworth, and Simon Alexander. Most of all, however, we are grateful to the extensive list of scholars from around the world who have agreed to contribute, and who make this series what it is.

Meegan Kennedy and Piers J. Hale

GENERAL INTRODUCTION TO THE COLLECTION

Scientific Methodology in Nineteenth-Century Britain

At the close of the nineteenth century, the atmosphere of recent scientific achievement was, at least for some, palpable enough to cast doubt on the future of scientific study. The physicist A. A. Michelson, speaking in 1894, claimed that "it seems probable that most of the grand underlying principles [in physics] have been firmly established and that further advances are to be sought chiefly in [their] rigorous application." Former physics professor and future president of the American Academy for the Advancement of Science T. C. Mendenhall wrote a few years earlier that "it is safe now to say what is possible and what is impossible. . . . [I]t may safely be affirmed that in electricity the past hundred years is not likely to be duplicated in the next."[1] While many historical evaluations of these kinds of claims oscillate between mockery of their short-sightedness and the self-assurance of historical hindsight, I want to focus on a slightly different aspect of such assertions, particularly evident in the way that Mendenhall formulates his version of the assessment. For if these authors thought that we can indeed be certain that future progress in the study of electricity will not parallel that of the past hundred years, this implies not just confidence in the scientific results which we already possess – it implies a refined understanding of scientific method. What is amenable to scientific investigation? What results may we reasonably expect to obtain, and what lies (or will forever lie) beyond the limits of human knowledge? To answer these kinds of questions, we need a nuanced understanding of scientific methodology and epistemology.

It is the aim of this collection to demonstrate that – whatever we might think about the merits of the conclusions that these thinkers went on to draw – it was not at all unreasonable to think that the scientific community was, indeed, in possession of just such an understanding of its own methods and limits. In parallel with the massive strides in a host of scientific domains that the nineteenth century brought – electricity and thermodynamics, evolution and uniformitarian geology, social surveys and psychological experiments – we find an explosion of careful, considered, sophisticated thinking about just what it is that the sciences are in fact doing, how their methods can be expected to produce reliable knowledge of the external world, and in what kinds of circumstances they could potentially fail.

While some of this body of thought was developed as abstract theorizing – as we will see especially in the first volume, the century produced numerous grand methodological treatises – much more of it, arguably, was built in tight dialogue with scientific practice, as scientists found themselves in need of justifications for the very sorts of investigations that they wanted to pursue. A review of these works, then, also doubles as a tour of some of the highlights of and biggest names in nineteenth-century science.

One question naturally arises upon opening a collection dedicated to scientific methodology in the nineteenth century: in what sense, exactly, are we going to consider "methodology"? The response to that question in these volumes will be somewhat unorthodox. To be sure, the 1800s saw vast refinements in technical, mechanical, and experimental methods, refinements that permitted the study of systems that had previously been wholly inaccessible, or with a precision that had been entirely inconceivable. The century began with Henry Cavendish (1731–1810) obtaining the first accurate measurement of the Newtonian gravitational constant,[2] and concluded with Theodor Boveri (1862–1915) and Walter Sutton's (1877–1916) experimental elucidation of the chromosomal theory of inheritance.[3] A collection of this size could easily be spent exploring myriad practical advancements like these and considering their impact on the future trajectory of the sciences. Such a focus on the objects of scientific practice would also resonate with recent historiographical trends toward the consideration of material culture, which has been central in the history of science especially since the 1990s.[4] This is, however, not the approach that I will take here. Exploring those experiments is of paramount importance, as well, for understanding the histories of the particular sciences that they impacted (in our two examples here, mechanics – or geodesy – and molecular biology). The descriptions of those experiments and the technical innovations that made them possible, then, is best saved for the other collections in this series that deal directly with the very sciences involved. As we will see, in a few places the development of particular tools (like statistical analysis or psychological tests) will occupy our attention. But this will not be the sense of "methodology" that shapes this volume.

My understanding of "methodology," in contrast, is perhaps one level more abstract than this. Throughout the four volumes that follow, we will trace changes brought in the nineteenth century not only in the content of scientific disciplines, and in the techniques we use to access the external world, but in the very idea of what it meant to be doing science. Entire domains of natural phenomena yielded to scientific analysis over this century, from studies of the fundamental structure of matter and energy, to scientific approaches to the evolution of the planet and the life on it, or the nature of human minds and societies. It is this vast array of advancements that this collection takes as its object. The very ways in which scientific knowledge was obtained, structured, legitimated, and debated would transform over the course of the 1800s to bring these domains within the purview of the scientific project. Of course, this places us squarely on the frontier between the history of science and the philosophy of science. In that sense, I take this collection to be a contribution to the study of the integrated history and philosophy of science, hopeful to produce work that is simultaneously valuable and important by the standards of both the history of science and the philosophy of science.[5] Determining whether or not it succeeds in doing so is an exercise left to the reader.

It is noteworthy that there is no general historiographical position on the material included here in the history of science community. Consider John Herschel's *Preliminary Discourse on Natural Philosophy* – one of the central texts of the first volume – as an object lesson. Lukas Verburgt summarizes his discussion of it in the recent *Cambridge Companion to John Herschel* by saying that there "exists little to no consensus on any aspect of the book, not even on what it is really about."[6] The volume of the *Cambridge History of Science* covering the physical sciences includes one chapter on "Theories of Scientific Method," which cites none of the authors we will discuss here (and next to nothing from the nineteenth century generally).[7] The volume on social sciences does better, with extensive coverage of the development of disciplinary structure and methodologies of a number of different sorts – particularly exemplary here is Stephen Turner's chapter on "Cause, Teleology, and Method," which cites, from among our authors, Comte, Kant, Mill, Reid, and Spencer.[8] (The volume on biological and earth sciences, surprisingly, includes no methodological chapter whatsoever.) To the extent that there is a tradition of discussion of scientific method present in the nineteenth century, it seems largely absent from the historiography of the period.

Part of the problem, as we will see, reflects the extremely diverse reading public for a work like Herschel's *Preliminary Discourse.* We know that numerous practicing scientists read it (including at crucial moments in the preparation of their own works), made reference to it, praised it, and even produced texts that seemed to follow its "best practices."[9] But what exactly they drew from it is never entirely clear, as direct citations are few and far between. We also know that many other, non-scientific audiences engaged with it as well, including general readers who were interested in improving their character by understanding how men of science thought and approached the world. So what exactly is this text? Is it a treatise in the philosophy of science? It certainly can be read that way, and we will see below that the history of philosophy has occasionally, though infrequently, considered it as such. Is it a lens that can be used to understand the works of other nineteenth-century authors? Assuredly; this has been profitably pursued, at the very least, for the cases of Charles Lyell and Darwin.[10] Is it a public text, to be placed in the tradition of the popularization of science?[11] It certainly could be – its presentations of recent work in geology and on theories of light would have been accessible to nearly any educated audience. Or could it even be read as a kind of "manual of conduct," a sort of nineteenth-century contribution to the "self-help" genre? Indeed; James Secord has forcefully made precisely this argument.[12]

Such a profusion of historical approaches is, to some degree, expected: not only are these authors writing outside of any particularly well-defined genre, they are also writing in a period during which their own professional role was rapidly changing. (We will return to the question of the term "scientist" itself, which nicely encapsulates this problem of professionalization.) This collection will run the gamut from authors who were self-financed and among the intellectual, cultural, and social elite, like Herschel, via those, such as Whewell, who came from at least more humble beginnings but quickly found themselves among the upper classes, to those like Faraday or Wallace who had to make a living through their scientific pursuits and lacked anything like an elite level of financial or social capital.[13] In addition to their content, then, the social and cultural context in which these works were created was far from uniform, an influence which we will explore to some extent, though doubtless not enough, in the commentaries that follow.

If one potential reading of these texts is philosophical, then there is also a historiographical approach to them that we can trace across histories of philosophy. At the very beginning of the century, contributing to the *Dictionary of Philosophy and Psychology* edited by James Mark Baldwin, Charles Saunders Peirce would write in the entry on "Methodology" that

> for the last three centuries thought has been conducted in laboratories, in the field, or otherwise in the face of the facts, while chairs of logic have been filled by men who breathe the atmosphere of the seminary. The consequence is that we can appeal to few works as showing what methodology ought to be.[14]

He does, however, cite as fruitful the very tradition that will appear here: Francis Bacon's *Novum Organum* and Jean Senebier's *L'art d'observer* make up the pre-nineteenth-century references, and across the nineteenth century he mentions, among others, Comte, Whewell, Mill, Jevons, Pearson, and Wundt. But the recognition of this tradition was short-lived, it seems. E. A. Burtt's *Metaphysical Foundations of Modern Science*, published in 1932, makes reference only to Mach and Whewell in its bibliography, and does not cite them in the main text.[15] In the middle of the century, Frederick Copleston's celebrated synthetic history of philosophy (published in the 1960s) briefly covers Darwin and Karl Pearson, and stops a bit longer at Herbert Spencer, but the highlights of the nineteenth century are clearly taken to be Jeremy Bentham and John Stuart Mill's utilitarianism (*not* Mill's epistemology), the idealism of F. H. Bradley and Bernard Bosanquet, and the roots of analytic philosophy.[16] The recent and much acclaimed *New History of Western Philosophy* by Anthony Kenny makes room for Darwin, and its section on modern epistemology includes Mill's *System of Logic* as a bastion of empiricism, but skips from Mill straight to Peirce.[17] The collection *Philosophy of Science: An Historical Anthology* comes closest to recognizing the historical arc that we will trace here, covering (from our authors) Herschel, Mill, Whewell, Lyell, Darwin, and Einstein before launching into contemporary works, but it is made conspicuous by its uniqueness in this regard.[18] In short, the history of philosophy also seems to have largely failed to recognize the coherence and importance of the works that we will explore here. Whether we consider the developments that traverse this collection to be part of the history of science, the philosophy of science, or some mix of the two, the fact remains that traditional work in both fields has not taken them to be a coherent object of study. One of the principal hopes of this collection is to promote such study in the future.

One further terminological clarification needs to be made explicit before we conclude this general introduction. How should we refer to the authors that we will encounter across these four volumes? At the beginning of the century, all would have considered themselves philosophers, or natural philosophers. By the century's end, many would have considered themselves to be professional scientists. This is not an anodyne question; in fact, it leads us to the heart of difficult problems about the status of science and its practitioners in the nineteenth century. As Sidney Ross puts it, "*scientist*, with its overtones of specialism and professionalism, was not in accord with the *persona* that the gifted amateur had of himself and his scientific pursuits."[19] An anonymous review from 1834, written, as it happens, by William Whewell, of Mary Somerville's *On the Connexion of the Physical Sciences* – we will see excerpts from both authors in what follows – was the first to introduce the term *scientist* in English. But Whewell backed away from it almost immediately. He describes the introduction of the term by "some ingenious gentleman" (who, Ross notes, is actually Whewell himself) at a meeting of the British Association for the Advancement of Science, "by analogy with *artist*," following by saying that "this was not generally palatable."[20] Michael Faraday would, to his dying day, accept neither *scientist* nor (the equally newly coined) *physicist*, insisting that he remained a *philosopher*. William Thomson preferred *naturalist*, which to our ear evokes only *natural historian*, a favored epithet of those like Jean-Baptiste Lamarck and Charles Darwin. Some British authors would persist in (derisively) calling *scientist* an American import, where it had taken off more quickly, given the lesser American reticence to associate with the "technical" and the "practical."[21]

In what follows I will, for ease of exposition, refer to all our authors as 'scientists.' The alternative – scrupulously keeping track of which authors preferred which labels (Alfred Russel Wallace: yes, Faraday and Thomson: no . . .) – will run aground when we are forced to deal with those authors who never expressed an opinion on the matter, but could have done so (to my knowledge, e.g., Darwin never did). It also would decrease readability for only a very minor gain in historical accuracy. I hope that Faraday, Thomson, and other detractors will be able to forgive the indiscretion; if not, they may ascribe it to my being of American extraction.

The four volumes of this collection are structured as follows. The first volume considers systematic approaches to the philosophy of natural science – not an invention of the nineteenth century, but, for the first time, a format which found itself at the center of a vibrant and large community of authors who were in dialogue with one another and saw themselves responding to a shared collection of methodological and meta-methodological concerns. (One might say that, for the first time, the philosophy of science began to find itself doing "normal science" in Thomas Kuhn's sense.) Each subsequent volume picks up a different domain of the sciences – in the second, the study of deep time in geology and evolutionary theory, in the third, the social and psychological sciences, and in the fourth, the physical sciences. Each volume, in turn, is subdivided into thematic parts, and each part is provided with its own short introduction describing the unifying thread connecting the excerpts which it contains. Each text, then, is preceded by a short, descriptive headnote – depending on the text, this might provide situating context, draw out relationships with other authors, or pick out particularly important or challenging passages for interpretation. Questions of demarcation are, to be sure, always difficult (do geology's connections with physics make it a physical science, say, or does its evidential usefulness for and epistemic similarity with evolution justify its placement alongside the life sciences?), but I think the choices made here are defensible ones, especially when considered at the level of methodological and evidentiary commitments.

I have many people to thank for their help exploring figures in these volumes who are outside my usual scholarly range. In particular, discussions with J. T. Burman, Steve Case, Christopher Green, Ryan Nichols, Trevor Pearce, Greg Radick, and Pieter Thyssen were instrumental in my approach to these texts. Among these, Christopher Green, Ryan Nichols, and Pieter Thyssen went the extra mile and checked some of the headnotes for accuracy. There are doubtless mistakes, both of the small factual and the more significant interpretational variety. I hope they will be few in number, and they are, of course, to be attributed to me alone. I also have to thank the series editors, Piers Hale and Meegan Kennedy, and my contacts at Routledge, Simon Alexander and Rachel Douglas, who were all kind enough to have invited me to prepare this project, and compassionate enough to have allowed me to take significantly longer in preparing it than I had anticipated. Lastly, I am as always grateful for the support of my family – especially my mother, Cynthia Pence, whose editing prowess always makes me seem significantly more eloquent than I actually am.

Notes

1 Both quoted in L. Badash, 'The Completeness of Nineteenth-Century Science,' *Isis* 63:1 (1972), pp. 48–58.

2 R. McCormmach, 'Mr. Cavendish Weighs the World,' *Proceedings of the American Philosophical Society* 142:3 (1998), pp. 355–66.

3 M. D. Laubichler and E. H. Davidson, 'Boveri's Long Experiment: Sea Urchin Merogones and the Establishment of the Role of Nuclear Chromosomes in Development,' *Developmental Biology* 314:1 (2008), pp. 1–11.

4 See, e.g., A. Appadurai (ed.), *The Social Life of Things: Commodities in Cultural Perspective* (Cambridge: Cambridge University Press, 1986).

5 T. Arabatzis and D. Howard, 'Introduction: Integrated History and Philosophy of Science in Practice,' *Studies in History and Philosophy of Science* 50 (2014), pp. 1–3.

6 L. M. Verburgt, 'Herschel's Philosophy of Science: The *Preliminary Discourse on the Study of Natural Philosophy* and Beyond,' in S. Case and L. M. Verburgt (eds), *The*

Cambridge Companion to John Herschel (Cambridge: Cambridge University Press, 2024), pp. 99–126.

7 N. Cartwright, S. Psillos, and H. Chang, 'Theories of Scientific Method,' in M. J. Nye (ed.), *The Cambridge History of Science, Volume 5: The Modern Physical and Mathematical Sciences* (Cambridge: Cambridge University Press, 2002), pp. 19–35.

8 S. Turner, 'Cause, Teleology, and Method,' in T. M. Porter and D. Ross (eds), *The Cambridge History of Science, Volume 7: The Modern Social Sciences* (Cambridge: Cambridge University Press, 2003), pp. 57–70.

9 C. H. Pence, 'John Herschel's Methodology in the Scientific Community,' in S. Case and L. M. Verburgt (eds), *The Cambridge Companion to John Herschel* (Cambridge: Cambridge University Press, 2024), pp. 257–79.

10 These are summarized in Pence, 'Herschel's Methodology.'

11 E.g., P. J. Bowler, *Science for All: The Popularization of Science in Early Twentieth-Century Britain* (Chicago: University of Chicago Press, 2009).

12 J. A. Secord, *Visions of Science: Books and Readers at the Dawn of the Victorian Age* (Chicago: University of Chicago Press, 2014), pp. 80–106.

13 And financial capital itself meant different things at different times across the century; see, e.g., M. J. S. Hodge, 'Capitalist Contexts for Darwinian Theory: Land, Finance, Industry, and Empire,' *Journal of the History of Biology* 42:3 (2009), pp. 399–416.

14 C. S. Peirce, 'Methodology,' in J. M. Baldwin (ed.), *Dictionary of Philosophy and Psychology* (New York: Macmillan, 1901), at *https://psychclassics.yorku.ca/Baldwin/Dictionary/* [accessed 13 January 2025].

15 E. A. Burtt, *The Metaphysical Foundations of Modern Science*, 2nd edn (New York: Doubleday, 1932).

16 F. Copleston, *A History of Philosophy, Volume VIII: Modern Philosophy: Empiricism, Idealism, and Pragmatism in Britain and America* (New York: Doubleday, 1966).

17 A. Kenny, *A New History of Western Philosophy, Volume IV: Philosophy in the Modern World* (Oxford: Clarendon Press, 2007).

18 T. McGrew, M. Alspector-Kelly, and F. Allhoff (eds), *Philosophy of Science: An Historical Anthology* (Malden, MA: Wiley-Blackwell, 2009).

19 S. Ross, *Nineteenth-Century Attitudes: Men of Science* (Dordrecht: Kluwer, 1991), pp. 1–2.

20 Quoted in S. Ross, *Nineteenth-Century Attitudes*, p. 9.

21 S. Ross, *Nineteenth-Century Attitudes*, pp. 11–2.

INTRODUCTION TO VOLUME I

Building Philosophical Systems

The philosophy of science by no means began in the nineteenth century. Ancient authors in nearly every philosophical tradition, from Aristotle to Nāgārjuna, theorized about the nature and function of our knowledge of the external world and the methodology by which we might attain it. It is nevertheless undeniable that the mid- to late-nineteenth century saw a period of systematic theorizing about the sciences that was unprecedented, perhaps only matched by the institutionalization of the philosophy of science as a discipline a hundred years later. The goal of this first volume is to present a cross-section of this work, as well as some of the scientific and philosophical background necessary to understand the context in which it was created.

In terms of scientific methodology, the century begins in the long shadows of Francis Bacon and Isaac Newton. Newtonian mechanics was, far and away, the paradigmatic example of what good scientific theorizing was supposed to resemble. But the legacies of Newton and Bacon are complex. Bacon's invective against the "idols of the mind," the (bad) habits of thought that take us away from the truth, had cautioned us in particular against the damaging influence of theoretical or speculative reasoning, epitomized for him by Aristotle (and, later, exemplified for Newton by Descartes), which could lead us to lose sight of the observed data that ought to lie at the basis of any scientific inference. At the same time, Bacon's detailed methodology for uncovering the cause of a phenomenon (involving the construction of tables of the presence or absence of the phenomenon and each of the causes that we suspect to be at work) was highly constraining, and very few, if any, practicing scientists had actually implemented it in their work. Bacon thus stood as an exemplar, but a kind of symbolic exemplar who seemed not to be followed in the development of any genuine scientific advances.

Most troublesomely of all, it seems that Newton had not argued in a particularly Baconian manner – Bacon's name is not mentioned in the *Principia*, and is only briefly referenced in the *Opticks* as a representative of the "new science" of the Renaissance, along with Boyle and Galileo. Newton did apparently share Bacon's worries about theoretical reasoning, famously arguing that he did not "feign hypotheses," a phrase that will be the subject of much more discussion in the pages that follow. At the same time, the obvious methodological difficulty facing Newtonian mechanics (as well as Newton's optics, along with various attempts by others to build scientific theories in a broadly "Newtonian" fashion) was the inference to the existence of forces like that of gravitation. The notion of force is not Newton's own invention, but he had indeed placed forces as perhaps the central concept of his theory.[1] By the time the *Principia* reaches its conclusion, he had introduced an entire taxonomy of forces – but force itself remains, in the end, "a concept given a priori, intuitively, and ultimately in analogy to human muscular force."[2] While force as something directly transmitted by impact, for instance, might have been a relatively easy concept to comprehend, things become much more difficult when we turn to gravity. Gravity is constantly in action, universal, capable of acting instantaneously at a distance, and does not seem to be transmitted by any kind of medium that can be observed or experienced directly. Newton, in the queries to the *Opticks*, speculated about the possibility of an "aetherial medium" that would exist in otherwise apparently empty space, but ultimately leaves the question open.[3] This, as we now know from his correspondence and his (unpublished during his lifetime) *De Gravitatione*,[4] was one of many times that Newton entertained what might have otherwise been considered to be highly speculative "hypotheses." We thus are left with no clear orientation toward hypothesizing or theorizing in the works of either of the methodological luminaries whose influence reigned in 1800 in Britain. Speaking of Herschel and Whewell's philosophies of science, about which more in a moment, David Wilson summarized this common point of departure, writing that it "drew its matter from Newtonian mechanics and astronomy, its form from the legacy of Newton's natural theology and rules of reasoning."[5]

At first blush, this might seem like a relatively limited squabble over how seriously we should take theoretical knowledge and how much space we should give to our speculative flights of fancy. That, in and of itself, would be an interesting but perhaps minor dispute over proper scientific method. But as we will see, the extent of the questions that it raises is much larger than this. First, we can consider debate over the nature, or even the possibility, of causal knowledge. A traditional view of the aims of science would have it that science tells us about the causes of phenomena, or, cast into slightly more modern language, that science offers us causal explanations of those phenomena. It is clear that our position with respect to "speculation" is crucial for answering this question. Just what does it mean to have actually offered a causal explanation for a phenomenon? With Descartes, do we need to have reduced it to the transmission of mechanical force by impact? Or would proposing a force suffice for a causal explanation, if we have supplied sufficient evidence for its existence and action? If so, what are those criteria? If not, what would permit us to say that such an explanation is not legitimate? We arrive quickly at a need not just for a limited opinion about the nature of forces, but for a full understanding of what kinds of explanations are sought by the sciences, and what kinds of evidence it takes to confirm or refute those explanations as scientifically suitable.

Second, the unease surrounding "hypothesis" and "speculation" as we saw it expressed by Bacon and Newton leads us to consider the relationship between our observations and our theories. Bacon's answer to this question is often presented as a kind of naive inductivism, on which we accumulate empirical data until we feel confident enough to form from them a simplistic generalization. This is too reductive a view of Bacon's position, however; his picture of the interplay of the empirical and the rational elements of scientific activity leaves a clear place for theorizing.[6] At the same time, there was fairly widespread agreement that what we might call "excessive" theorizing was a bad thing; Newton was on to something when he cautioned us against hypotheses. We thus stand in need of rules, or at least guidelines, that can tell us when and how we should turn to the construction of scientific theories. Can we observe phenomena "directly" at all, that is, in the absence of theoretical categories in which to classify and understand those observations? How should we respond to skeptical arguments that we cannot directly see causal relationships in the world around us, or the unobservable entities that science proposes? The relationship of *a priori* knowledge with our observations also stands in need of clarification. On the one hand, we see over this period the increasing importance of mathematical knowledge, which would become applicable not only within the physical sciences, but also increasingly essential across the scientific enterprise. This knowledge is itself not directly drawn from our observations, and this could be taken to be a threat (and was, by authors like John Stuart Mill) to the "observational" legitimacy of the theories that result. The same is true for the use of philosophical reflection. Some (especially William Whewell) argued that metaphysical analysis, far from being speculative or theoretical, was essential in order to give our scientific knowledge a firm foundation. Others would reject any such philosophical approaches, perhaps (with the positivists) consigning them to a prior, undeveloped stage of scientific reasoning.

The majority of the works considering these questions were published in the first half of the nineteenth century. Their examples were largely drawn from physics and chemistry, with the latter representing, as it were, the cutting edge of scientific inquiry. But as the years wore on, Darwin's theory of evolution (about which more in the next volume) and nascent theories of the human and mental sciences (which we will see in Volume 3) would also put pressure on these procedures of causal reasoning. In these fields, where data are even harder to come by, we could find ourselves unable to apply the tools that we had developed in the physical sciences.

This collection of difficult, open problems gave rise to a flourishing production of major works in methodology and philosophy of science, dominated by the important (and, in the latter two cases, immense) works of Sir John Herschel, William Whewell, and John Stuart Mill. These were arguably the first attempts at a systematic treatment of these subjects since Bacon's *Novum Organum*, and were eagerly devoured by the scientific establishment of the day. To be sure, dedicated works on scientific methodology are not the only place where this kind of thought was presented during this period. The following three volumes of this collection will, as noted previously, consider the ways in which different sciences take on these questions in their own ways. But the impact of these seminal works was tremendous. They were also widely read among the general public. As Victorian Britain took science to be the gold standard for knowledge, its citizens equally took scientific methodology to be the gold standard for the pursuit of truth. Learning about that pursuit of truth is thus personally, or even morally, edifying, the kind of reflection appropriate for men (and, as we will see, women, as the scientific establishment began to lower barriers to women's entry) of culture. By the second half of the century, one could count on every "scientist" being familiar with at least the broad outlines of these books, even if that knowledge wasn't immediately obvious in the context of their own work.

Mentioning these three authors – Herschel, Whewell, and Mill – is a natural transition to the social aspect of this endeavor. As we will see, the scientific establishment of nineteenth-century Britain was a close-knit one, united by central universities (Whewell, for instance, was one of the most powerful figures at Cambridge), by key institutions (such as the Royal Society and the British Association; Whewell and Herschel each served as president of the latter), and by a tight web of social links that bound the vast majority of the authors that will be present across these four volumes. Interconnections between the material treated by them are frequent – for instance, Herschel presents a version of the "consilience of inductions," a concept usually ascribed to Whewell, a number of years before Whewell would publish it, and all three of these figures would present similar compilations of rules of thumb for causal inference from observational and experimental data. Determining "priority" is almost impossible (and, in any case, not particularly interesting) in this context. Better, by contrast, to look at their social connections as evidence of a shared intellectual enterprise, a community approach to questions of sound methodology that permeated scientific practice. While the selections in this volume may be relatively light on scientific detail – which will amply return in the following three volumes – as a result of this community, they form the theoretical context for many of the scientific works that would reshape our understanding of the natural world across the nineteenth century.

To close this introduction, let us briefly turn to the answers that these authors gave to each of these two major sets of questions. All three of Herschel, Whewell, and Mill took the construction of causal knowledge very seriously. Herschel and Mill both found it essential to lay down rules for deriving causal connections from observations, and they each carefully explored under what circumstances such a conclusion would be acceptable. The most lasting impact in this area was Mill's; his series of five methods for moving from experiment to causal explanation is now stock in trade in many presentations of experimental design in the sciences. For him, the very subject of logic was inference from some true claim about the external world to another such claim about the external world – all inference is fundamentally inductive. Herschel, for his part, emphasized the process of verification of a purported cause – proposing any number of speculative causes is, he argued, relatively easy, while ensuring that the cause we have selected is really responsible for the phenomena that we claim it is poses the real challenge. But both were motivated by the idea that in spite of the perhaps tenuous nature of the inferences made to arrive there, Newtonian mechanics had given us at least something approaching causal understanding of the world, and any methodological dictum worth its salt needed to recognize that fact.

For Whewell, things are a bit more complex. He also wanted to defend the possibility of causal knowledge, but he sought a deeper grounding for that knowledge than its successful track record or its having passed a number of observational or experimental tests. Following in the footsteps of Kant, for Whewell, causation is fundamentally a philosophical concept, one which is given an important part of its content by *a priori* philosophical arguments – for instance, the argument that no event can be uncaused. For Whewell, then, the challenge is to bring our observational and experimental work into dialogue with fundamental metaphysical convictions. This brings us naturally to the second major question that was just raised: the relationship between our theories and our observations. For Whewell, this becomes an especially difficult problem: our theoretical work sits at a fulcrum between philosophical argumentation concerning concepts like causation or force and our empirical observations or experiments that involve those very same ideas. We thus have a particularly acute need of careful philosophical analysis of the sciences, because it is precisely this that can tell us when our scientific work has demonstrated to us a necessary, metaphysical truth about the world. Many of his colleagues remained unconvinced that such a demonstration was even possible.

On this second question, Mill was also fairly extreme, albeit in the opposite sense. As we have already seen, for Mill there can be no questioning the supremacy of observation over theory, even in the domain of apparently necessary truths such as those of mathematics or geometry. It is notable, however, that this answer to the relationship between theory and observation makes for some tension with his emphasis on the discovery of causal knowledge. Mill clearly has no use for a generalized skcpticism about theory; we simply need to structure our observations in the appropriate manner to be sure that we are indeed discovering theories of genuine causes, to follow the correct logic of observation. In this, one might see in Mill a kind of precursor to modern views of causal inference, a parallel that has been considered by a number of contemporary philosophers of science.[7] But for Mill, the fundamental role of theory, as is made clear by the methods that bear his name, is to tease apart the vast, complex causal nexus that surrounds every phenomenon and offer us a refined understanding of the various, often independent influences that give rise to the world around us.

The observational focus of Mill's thought parallels that of the positivists, who would go even farther, placing causal knowledge entirely beyond our reach. For them, our theories are only useful tools that enable us to describe the observations that we have already obtained and to predict future observations of similar kinds in similar conditions. Beyond this, the content of our theories is not to be trusted; that way lies a return to the "metaphysical" stage of scientific theorizing. No matter what philosophical argumentation we might produce, what we might observe, or what procedure we might use to verify it, we should not pretend that our theories are anything more; any such appearance is an illusion best set aside to ensure the progress of science.

Herschel, finally, is resolutely centrist. For him, theoretical knowledge does, at least at times, provide us with genuine insight into the natural world. Particularly important for him here is the wave theory of light, which he believed offered a very important advance in understanding, but which he knew lacked the kind of evidence needed to satisfy anyone with strong empiricist leanings. Mill, for instance, argued that even the successful predictions produced by the wave theory of light did not offer us any proof of the existence of the ether, the medium that was supposed to carry those waves. Herschel did not want to give in so easily. For him, theoretical knowledge can indeed unveil the underlying fundamentals of the universe, but without the neo-Kantian support to which Whewell would turn. How exactly to make this position coherent is among Herschel's most important challenges; his response, turning to practices of verification, unsurprisingly, wasn't a basis strong enough for Whewell's taste, however much Whewell argued that his philosophy, too, was derived directly from the historical practice of the sciences. As Wilson notes, while "Whewell and Herschel agreed that the laws were just as certainly true as mathematical propositions," the relationship between "internal experience" and "external experience" that this certainty required for Herschel is a difficult one.[8]

As we can see, while there are similarities between the thought of the figures that we will encounter in this volume, spurred no doubt at least in part by the extensive personal and professional relationships that knitted together the nineteenth-century British scientific community, there are also genuine and important differences of opinion at work here. These thinkers were attempting to respond to a world that contained not only Newtonian mechanics as, it seemed, a kind of finished monument to scientific success, but also theories whose legitimacy was not yet subject to wide agreement, novel methodology, and newly proposed theoretical entities. They were each, as well, highly credentialed practicing scientists, experts whose right to weigh in on such matters was beyond dispute. It is for this reason that these nineteenth-century philosophical and methodological works have proven to be such rich sources of insight.

Notes

1 M. Jammer, *Concepts of Force: A Study in the Foundations of Dynamics* (Cambridge, MA: Harvard University Press, 1957).
2 M. Jammer, *Concepts*, on p. 124.
3 I. Newton, *Opticks* (New York: Dover, 1952), pp. 350–3.
4 See excerpts published in A. Janiak (ed.), *Isaac Newton: Philosophical Writings* (Cambridge: Cambridge University Press, 2004).
5 D. B. Wilson, 'Herschel and Whewell's Version of Newtonianism,' *Journal of the History of Ideas* 35:1 (1974), pp. 79–97, on p. 80.
6 J. Klein, 'Francis Bacon,' *Stanford Encyclopedia of Philosophy* (Fall 2020), *https://plato.stanford.edu/archives/fall2020/entries/francis-bacon/* [accessed 21 June 2024].

7 For a subtle approach along these lines, and references to further literature, see R. Scholl, 'Inference to the Best Explanation in the Catch-22: How Much Autonomy for Mill's Method of Difference?,' *European Journal for Philosophy of Science* 5 (2015), pp. 89–110.
8 D. B. Wilson, 'Herschel and Whewell,' p. 82.

Part 1

SETTING THE STAGE

This nineteenth-century revolution in scientific methodology did not arise *ex nihilo*. In particular, the colossal impact of Newton's *Principia*, which revolutionized physics in the middle of the eighteenth century, had by no means been fully digested by the scientific and philosophical communities (or, perhaps better, the community of natural philosophers) as that century came to a close. Newton's works bequeathed those scholars a host of interrelated and extremely difficult issues. We have already seen the question of the justification of hypotheses, which is rendered all the more poignant by Newton's apparent injunction that we are not to "feign" them (or "frame," "pose," or however one prefers to render the Latin *fingo*). Nonetheless, Newton was himself a prolific generator of hypotheses, and further hypothetical works over the course of the eighteenth century in fields like astronomy, electrical theory, or chemistry seemed to be yielding fruit.

Further, Newton had offered us something that seemed to many to be the paragon of scientific methodology and a metaphysics to go along with it: we observe the behaviors of the particles out of which bodies are composed, we comprehend the fundamental forces that operate upon them, and we learn how those forces combine to produce different sorts of behaviors at different levels of organization. But the application of this paradigm to other areas had proven difficult. Just what are the "forces" of Newtonian chemistry, or biology, or medicine supposed to be?[1]

Finally, how are we supposed to deal with the foundations of these sciences? Newton helped himself to concepts like motion, action, reaction, force, cause, and effect. Are these to be taken as fundamental or primitive? Or should we analyze them further? Are they to be derived via *a priori* metaphysical reasoning, or are they a simple matter of common sense?

All of these questions (and others that I lack the space to treat here) created an environment in which working on questions of scientific methodology – a discipline that we would later come to call the philosophy of science, though I tend to avoid that label in order to avoid anachronism – was a vital and important part not just of more "abstract" or "theoretical" branches of natural philosophy, but of works by more applied, experimental researchers as well, the class of natural philosophers who would soon begin to consider, following William Whewell's advice, referring to themselves as "scientists."

DOI: 10.4324/9781003263814-1

This first part of the volume, then, is intended to set this stage: to show what kinds of problems would have motivated the mid-nineteenth-century explosion of work on scientific methodology and made the need for such work seem inescapable.

Note

1 On the particularly interesting case of Newtonian medicine, see T. M. Brown, 'Medicine in the Shadow of the *Principia*,' *Journal of the History of Ideas* 48:4 (1987), pp. 629–48.

Editorial Headnote

1. Isaac Newton, "Scholium," from *Principia Mathematica*, tr. Andrew Motte (1st edn 1687, 3rd edn 1726, tr. 1729, edn reproduced here 1803)

No discussion of scientific methodology in nineteenth-century Britain could begin anywhere other than the work of Isaac Newton (1642–1727). Just what it meant to do science as this century began was to do what Newton (and perhaps, as will be discussed in more detail later in this volume, Bacon) had done. As the decades advanced and further discoveries accumulated, one of the most important aspects of contemporary thought about how to do science would be debate over just what this methodological legacy amounted to. We begin, therefore, with two texts from Newton which can help us to see the reasons this might be a difficult question.

The first is the Scholium to Newton's *Principia*. In this short text, which we might now describe as a work in the metaphysics of science, Newton wrestles with the problem of defining the concepts upon which his theory of motion fundamentally rests: absolute and relative concepts of space, motion, and time, as well as the relations between them. As Newton clearly notes, his theory is defined in terms of "absolute, true, and mathematical" notions of time and space – far different from the meanings that we might ascribe to these terms in everyday use. True motion, that which Newton's work is intended to describe, is motion with respect to absolute space. But the parts of absolute space "cannot be seen, or distinguished from one another by our senses" (p. 7), because our only access to space comes via the positions and movements of bodies within it. Following the example of Galileo, Newton notes that experiments carried out on a stably moving ship, or on the constantly moving Earth, don't give detectably different results (p. 6); it is therefore "impossible to know, from the [relative] position of bodies to one another in our regions, whether any of these do keep the same position [relative] to" some hypothetical body that is at true, absolute rest.

If this is the case, then why postulate absolute space in the first place? We do see, Newton argues, some glimpses of truly absolute motion. In an example that would go on to no small philosophical fame, he notes that "all the parts of revolving bodies endeavour to recede from the axis of motion" (p. 7). Imagine spinning a bucket full of water. At first, the bucket spins around the water and the water's surface is flat. After a while, the rotation is transferred to the water, and it starts to be pulled toward the edges of the bucket, its surface taking on a concave shape. Both before and after the shape change, the water has the same relationship to the bucket. The difference, Newton claims, could be seen as due to the water's rotating in absolute space (p. 8).[1] This argues for one "real circular motion of any one revolving body, corresponding to only one power of endeavouring to recede from its axis of motion, as its proper and adequate effect" (p. 9).

Much the same holds for time. Newton's absolute time "from its own nature flows equably and without regard to anything external" (p. 5); our relative measures of that time are based on motions, like those of the Earth. We know that the concepts of absolute and relative time are distinct precisely because relative measures of time need to be corrected with respect to one another – for instance, we must use astronomical observations to correct for the unequal length of the day compared with other measures like those from a pendulum clock (p. 6).

The implicit opponent to this view (rendered more explicit in other parts of the *Principia*) is, of course, René Descartes (1596–1650). For Descartes, space can be entirely identified with the relative position of bodies, as the world is made of continuous matter with no gaps (i.e., no vacuum). Newton believes that the examples that he presents here clearly refute the Cartesian view of space, as he has identified effects of motion that are incompatible with motion's being a purely relative concept. A Cartesian, in turn, could respond that Newton's idea of absolute space, divorced as it is from any possible interaction with our senses, is metaphysically incoherent. We see here the first moves in a philosophical controversy about the underlying nature of space and time that continues today.[2]

Importantly for our purposes, we also see a debate over how scientific methodology ought to proceed. Above and beyond Newton's examples of particular motions that favor one interpretation over another, what kinds of inferences could ground our belief in concepts like absolute space and time? Are these kinds of concepts the sort of thing about which science can and should teach us? What is the relationship between science – or, here, natural philosophy – and metaphysics? Newton does not give us clear answers to these questions. While he puts much emphasis on the examples of motions that he provides, which could lead one to assume that it is experiment that is primary in these cases, he notes that it remains possible to consider the rotations he discusses as, themselves, relative rotation with respect to something else (such as the distant stars). The available data, that is, will still underdetermine the conceptual foundations of our physical theory; they will remain unable to fully resolve the interpretative question that Newton has posed. How, then, should a careful scientist proceed when confronted with such a case?

Notes

1 See also T. Maudlin, 'Buckets of Water and Waves of Space: Why Spacetime Is Probably a Substance,' *Philosophy of Science*, 60:2 (1999), pp. 183–203.

2 N. Huggett, *Space from Zeno to Einstein* (Cambridge, MA: MIT Press, 1999).

1

SCHOLIUM FROM *PRINCIPIA MATHEMATICA*

Isaac Newton

Source: Isaac Newton, "Scholium," from *Principia Mathematica,* tr. Andrew Motte (1803 [1726, tr. 1729]), pp. 1:6–1:14.

Hitherto I have laid down the definitions of such words as are less known, and explained the sense in which I would have them to be understood in the following discourse. I do not define time, space, place and motion, as being well known to all. Only I must observe, that the vulgar conceive those quantities under no other notions but from the relation they bear to sensible objects. And thence arise certain prejudices, for the removing of which, it will be convenient to distinguish them into absolute and relative, true and apparent, mathematical and common.

I. Absolute, true, and mathematical time, of itself, and from its own nature flows equably without regard to any thing external, and by another name is called duration: relative, apparent, and common time, is some sensible and external (whether accurate or unequable) measure of duration by the means of motion, which is commonly used instead of true time; such as an hour, a day, a month, a year.

II. Absolute space, in its own nature, without regard to any thing external, remains always similar and immovable. Relative space is some moveable dimension or measure of the absolute spaces; which our senses determine by its position to bodies; and which is vulgarly taken for immovable space; such is the dimension of a subterraneous, an æreal, or celestial space, determined by its position in respect of the earth. Absolute and relative space, are the same in figure and magnitude; but they do not remain always numerically the same. For if the earth, for instance, moves, a space of our air, which relatively and in respect of the earth remains always the same, will at one time be one part of the absolute space into which the air passes; at another time it will be another part of the same, and so, absolutely understood, it will be perpetually mutable.

III. Place is a part of space which a body takes up, and is according to the space, either absolute or relative. I say, a part of space; not the situation, nor the external surface of the body. For the places of equal solids, are always equal; but their superficies, by reason of their dissimilar figures, are often unequal. Positions properly have no quantity, nor are they so much the places themselves, as the properties

DOI: 10.4324/9781003263814-2

of places. The motion of the whole is the same thing with the sum of the motions of the parts; that is, the translation of the whole, out of its place, is the same thing with the sum of the translations of the parts out of their places; and therefore the place of the whole, is the same thing with the sum of the places of the parts, and for that reason, it is internal, and in the whole body.

IV. Absolute motion, is the translation of a body from one absolute place into another; and relative motion, the translation from one relative place into another. Thus in a ship under sail, the relative place of a body is that part of the ship which the body possesses; or that part of its cavity which the body fills, and which therefore moves together with the ship: and relative rest, is the continuance of the body in the same part of the ship, or of its cavity. But real, absolute rest, is the continuance of the body in the same part of that immovable space, in which the ship itself, its cavity, and all that it contains, is moved. Wherefore, if the earth is really at rest, the body, which relatively rests in the ship, will really and absolutely move with the same velocity which the ship has on the earth. But if the earth also moves, the true and absolute motion of the body will arise, partly from the true motion of the earth, in immovable space; partly from the relative motion of the ship on the earth: and if the body moves also relatively in the ship; its true motion will arise, partly from the true motion of the earth, in immovable space, and partly from the relative motions as well of the ship on the earth, as of the body in the ship; and from these relative motions will arise the relative motion of the body on the earth. As if that part of the earth where the ship is, was truly moved toward the east, with a velocity of 10010 parts; while the ship itself with a fresh gale, and full sails, is carried towards the west, with a velocity expressed by 10 of those parts; but a sailor walks in the ship towards the east, with 1 part of the said velocity: then the sailor will be moved truly and absolutely in immovable space towards the east with a velocity of 10001 parts, and relatively on the earth towards the west, with a velocity of 9 of those parts.

Absolute time, in astronomy, is distinguished from relative, by the equation or correction of the vulgar time. For the natural days are truly unequal, though they are commonly considered as equal, and used for a measure of time: astronomers correct this inequality for their more accurate deducing of the celestial motions. It may be, that there is no such thing as an equable motion, whereby time may be accurately measured. All motions may be accelerated and retarded, but the true, or equable, progress of absolute time is liable to no change. The duration or perseverance of the existence of things remains the same, whether the motions are swift or slow, or none at all: and therefore it ought to be distinguished from what are only sensible measures thereof; and out of which we collect it, by means of the astronomical equation. The necessity of which equation, for determining the times of a phænomenon, is evinced as well from the experiments of the pendulum clock, as by eclipses of the satellites of *Jupiter*.

As the order of the parts of time is immutable, so also is the order of the parts of space. Suppose those parts to be moved out of their places, and they will be moved (if the expression may be allowed) out of themselves. For times and spaces are, as it were, the places as well of themselves as of all other things. All things are placed in time as to order of succession; and in space as to order of situation. It is from their essence or nature that they are places; and that the primary places of things should be moveable, is absurd. These are therefore the absolute places; and translations out of those places, are the only absolute motions.

But because the parts of space cannot be seen, or distinguished from one another by our senses, therefore in their stead we use sensible measures of them. For from the positions and distances of things from any body considered as immovable, we define all places: and then with respect to such places, we estimate all motions, considering bodies as transferred from some of those places into others. And so instead of absolute places and motions we use relative ones; and that without any inconvenience in common affairs: but in philosophical disquisitions, we ought to abstract from our senses, and consider things themselves, distinct from what are only sensible measures of them. For it may be that there is no body really at rest, to which the places and motions of others may be referred.

But we may distinguish rest and motion, absolute and relative, one from the other by their properties, causes and effects. It is a property of rest, that bodies really at rest do rest in respect of one another. And therefore as it is possible, that in the remote regions of the fixed stars, or perhaps far beyond them, there may be some body absolutely at rest; but impossible to know, from the position of bodies to one another in our regions, whether any of these do keep the same position to that remote body; it follows that absolute rest cannot be determined from the position of bodies in our regions.

It is a property of motion, that the parts, which retain given positions to their wholes, do partake of the motions of those wholes. For all the parts of revolving bodies endeavour to recede from the axis of motion; and the impetus of bodies moving forwards, arises from the joint impetus of all the parts. Therefore, if surrounding bodies are moved, those that are relatively at rest within them, will partake of their motion. Upon which account, the true and absolute motion of a body cannot be determined by the translation of it from those which only seem to rest: for the external bodies ought not only to appear at rest, but to be really at rest. For otherwise, all included bodies, beside their translation from near the surrounding ones, partake likewise of their true motions; and though that translation was not made they would not be really at rest, but only seem to be so. For the surrounding bodies stand in the like relation to the surrounded as the exterior part of a whole does to the interior, or as the shell does to the kernel; but, if the shell moves, the kernel will also move, as being part of the whole, without any removal from near the shell.

A property near akin to the preceding, is this, that if a place is moved, whatever is placed therein moves along with it; and therefore a body, which is moved from a place in motion, partakes also of the motion of its place. Upon which account all motions from places in motion, are no other than parts of entire and absolute motions; and every entire motion is composed out of the motion of the body out of its first place, and the motion of this place out of its place; and so on, until we come to some immovable place, as in the before-mentioned example of the sailor. Wherefore entire and absolute motions can be no otherwise determined than by immovable places; and for that reason I did before refer those absolute motions to immovable places, but relative ones to moveable places. Now no other places are immovable but those that, from infinity to infinity, do all retain the same given positions one to another; and upon this account must ever remain unmoved; and do thereby constitute, what I call, immovable space.

The causes by which true and relative motions are distinguished, one from the other, are the forces impressed upon bodies to generate motion. True motion is neither generated nor altered, but by some force impressed upon the body moved: but relative motion may be generated or altered without any force impressed upon the body. For it is sufficient only to impress some force on other bodies with which the former is compared, that by their giving way, that relation may be changed, in which the relative rest or motion of this other body did consist. Again, true motion suffers always some change from any force impressed upon the moving body; but relative motion does not necessarily undergo any change by such forces. For if the same forces are likewise impressed on those other bodies, with which the comparison is made, that the relative position may be preserved, then that condition will be preserved in which the relative motion consists. And therefore any relative motion may be changed when the true motion remains unaltered, and the relative may be preserved when the true suffers some change. Upon which accounts, true motion does by no means consist in such relations.

The effects which distinguish absolute from relative motion are, the forces of receding from the axis of circular motion. For there are no such forces in a circular motion purely relative, but in a true and absolute circular motion, they are greater or less, according to the quantity of the motion. If a vessel, hung by a long cord, is so often turned about that the cord is strongly twisted, then filled with water, and held at rest together with the water; after, by the sudden action of another force, it is whirled about the contrary way, and while the cord is untwisting itself, the vessel continues for some time in this motion; the surface of the water will at first be plain, as before the vessel began to move: but the vessel, by gradually communicating its motion to the water, will make it begin sensibly to revolve, and recede by little and little from the middle, and ascend to the sides of the vessel, forming itself into a concave figure (as I have experienced), and the swifter the motion becomes, the higher will the water rise, till at last, performing its revolutions in the same times with the vessel, it becomes relatively at rest in it. This ascent of the water shews its endeavour to recede from the axis of its motion; and the true and absolute circular motion of the water, which is here directly contrary to the relative, discovers itself, and may be measured by this endeavour. At first, when the relative motion of the water in the vessel was greatest, it produced no endeavour to recede from the axis: the water shewed no tendency to the circumference, nor any ascent towards the sides of the vessel, but remained of a plain surface, and therefore its true circular motion had not yet begun. But afterwards, when the relative motion of the water had decreased, the ascent thereof towards the sides of the vessel proved its endeavour to recede from the axis; and this endeavour shewed the real circular motion of the water perpetually increasing, till it had acquired its greatest quantity, when the water rested relatively in the vessel. And therefore this endeavour does not depend upon any translation of the water in respect of the ambient bodies, nor can true circular motion be defined by such translations. There is only one real circular motion of any one revolving body, corresponding to only one power of endeavouring to recede from its axis of motion, as its proper and adequate effect: but relative motions in one and the same body are innumerable, according to the various relations it bears to external bodies, and like other relations, are altogether destitute of any real effect, any otherwise than they may perhaps participate of that one only true motion. And therefore in their system who suppose that our heavens, revolving below the sphere of the fixed stars, carry the planets along with them; the several parts of those heavens, and the planets, which are indeed relatively at rest in their heavens, do yet really move. For they change their position one to another (which never happens to bodies truly at rest), and being carried together with their heavens, participate of their motions, and as parts of revolving wholes, endeavour to recede from the axis of their motions.

Wherefore relative quantities are not the quantities themselves, whose names they bear, but those sensible measures of them (either accurate or inaccurate) which are commonly used instead of the measured quantities themselves. And if the meaning of words is to be determined by their use, then by the names time, space, place and motion, their measures are properly to be understood; and the expression will be unusual, and purely mathematical, if the measured quantities themselves are meant. Upon which account, they do strain the sacred writings, who there interpret those words for the measured quantities. Nor do those less defile the purity of mathematical and philosophical truths, who confound real quantities themselves with their relations and vulgar measures.

It is indeed a matter of great difficulty to discover, and effectually to distinguish, the true motions of particular bodies from the apparent: because the parts of that immovable space in which those motions are performed, do by no means come under the observation of our senses. Yet the thing is not altogether desperate; for we have some arguments to guide us, partly from the apparent motions, which are the differences of the true motions; partly from the forces, which are the causes and effects of the true motions. For instance, if two globes, kept at a given distance one from the other by means of a cord that connects them, were revolved about their common centre of gravity, we might, from the tension of the cord, discover the endeavour of the globes to recede from the axis of their motion, and from thence we might compute the quantity of their circular motions. And then if any equal forces should be impressed at once on the alternate faces of the globes to augment or diminish their circular motions, from the increase or decrease of the tension of the cord, we might infer the increment or decrement of their motions; and thence would be found on what faces those forces ought to be impressed, that the motions of the globes might be most augmented; that is, we might discover their hindermost faces, or those which, in the circular motion, do follow. But the faces which follow being known, and consequently the opposite ones that precede, we should likewise know the determination of their motions. And thus we might find both the quantity and the determination of this circular motion, even in an immense vacuum, where there was nothing external or sensible with which the globes could be compared. But now, if in that space some remote bodies were placed that kept always a given position one to another, as the fixed stars do in our regions, we could not indeed determine from the relative translation of the globes among those bodies, whether the motion did belong to the globes or to the bodies. But if we observed the cord, and found that its tension was that very tension which the motions of the globes required, we might conclude the motion to be in the globes, and the bodies to be at rest; and then, lastly, from the translation of the globes among the bodies, we should find the determination of their motions. But how we are to collect the true motions from their causes, effects, and apparent differences; and, *vice versa,* how from the motions, either true or apparent, we may come to the knowledge of their causes and effects, shall be explained more at large in the following Tract. For to this end it was that I composed it.

Editorial Headnote

2. Isaac Newton, "Queries," from *Opticks* (1730)

Newton's masterpiece in optics, the *Opticks* of 1730, did much to lay the groundwork for thought about Newtonian "methodology" across the sciences in the eighteenth and nineteenth centuries. First and foremost, Newton turns here to a much wider array of scientific fields, especially in the series of "Queries" that were appended to the final editions of the book, and which extend far beyond optical phenomena to areas as disparate as what we might now call thermodynamics, chemistry, physiology, and perception, among others. Second, he does so in English, rather than in the Latin of the *Principia*, and without his name on the title page. It has been argued that these choices can give us a window into Newton's opinion concerning the status of the science of mechanics – which, as it was a more complete and mature field, merited axiomatic, Latin treatment – with respect to the other sciences, which were perhaps too "unfinished" to deserve such formalities.[1]

As Andrew Janiak has noted, the fact that the Queries are set off from the rest of the work proper and labeled explicitly as such can be read as an indicator of Newton's attitude toward the generation of novel, speculative scientific hypotheses. Far from an exceptionless dictum that we ought to "feign no hypotheses" (or, as he puts the matter in the *Opticks*, that "Hypotheses are not to be regarded in experimental Philosophy"), Newton, as we see here, "exhibited an avowed willingness to consider all manner of hypotheses," wanting only to make it clear that he does not accord them a status which they have not earned. A hypothesis, Janiak continues, "is not 'feigned' unless one adopts an unwarranted epistemic attitude toward it, for instance, if one asserts it to be the correct explanation of some documented natural phenomenon."[2]

As we see, Newton begins by considering how light might be similar to or different from the other kinds of forces that he has considered in the *Principia* – it seems to work at a distance, modulating its effect depending on the distance between source and target, just as gravity does. But as he continues speculating on how light might be integrated into a broader understanding of the natural world, he arrives at what almost seems to be a project for future natural philosophy. Not only light, but also gravity, magnetism, and electricity seem to work in this way, as the result of individual forces acting at a distance between the particular particles of bodies, taken individually. To be sure, composition of those particles into larger bodies will make the story more complex. But the basic outlines of a general metaphysics of science, along with a methodology for studying it, can be found within the last few pages of the "Queries." Break bodies down into their smallest possible component parts, study the forces of attraction and repulsion that might hold between those parts, then consider how the behavior of larger objects could re-emerge by composition once those forces are well understood. Newton sketches for us the outlines of how chemistry could be built on such a basis – as such, he writes, "Nature will be very conformable to her self and very simple."

Not only does such conformity and simplicity permit, at least potentially, a unified methodology for the study of every branch of the natural sciences, it also leaves the perfect amount of room for the action of a divine creator. For if the macroscopic coherence of bodies and their properties is due to a microscopic dance of particles and their forces, only the intelligent arrangement of those particles and forces by an all-knowing agent could possibly produce something like the steady functioning of plants and animals or the regular orbits of the solar system – "such a wonderful Uniformity in the Planetary System must," Newton writes, "be allowed the Effect of Choice." Generations of natural philosophers would go on to be inspired by these "Hints to be examin'd and improv'd by the farther Experiments and Observations of such as are inquisitive."

Notes

1 I. B. Cohen, 'The Case of the Missing Author: The Title Page of Newton's *Opticks* (1704), with Notes on the Title Page of Huygens's *Traité de la Lumière*,' in J. Z. Buchwald and I. B. Cohen (eds), *Isaac Newton's Natural Philosophy* (Cambridge, MA: MIT Press, 2001), pp. 15–45.

2 A. Janiak, 'Introduction,' in A. Janiak (ed.), *Isaac Newton: Philosophical Writings* (Cambridge: Cambridge University Press, 2004), pp. ix–xxxi, on p. xxvii.

2

QUERIES TO THE *OPTICKS*

Isaac Newton

Source: Isaac Newton, "Queries," from *Opticks* (1730), introduction, Qu. 1–7, 31.

When I made the foregoing Observations, I design'd to repeat most of them with more care and exactness, and to make some new ones for determining the manner how the Rays of Light are bent in their passage by Bodies, for making the Fringes of Colours with the dark lines between them. But I was then interrupted, and cannot now think of taking these things into farther Consideration. And since I have not finish'd this part of my Design, I shall conclude with proposing only some Queries, in order to a farther search to be made by others.

Query 1. Do not Bodies act upon Light at a distance, and by their action bend its Rays; and is not this action (*cæteris paribus*) strongest at the least distance?

Qu. 2. Do not the Rays which differ in Refrangibility differ also in Flexibity; and are they not by their different Inflexions separated from one another, so as after separation to make the Colours in the three Fringes above described? And after what manner are they inflected to make those Fringes?

Qu. 3. Are not the Rays of Light in passing by the edges and sides of Bodies, bent several times backwards and forwards, with a motion like that of an Eel? And do not the three Fringes of colour'd Light above-mention'd arise from three such bendings?

Qu. 4. Do not the Rays of Light which fall upon Bodies, and are reflected or refracted, begin to bend before they arrive at the Bodies; and are they not reflected, refracted, and inflected, by one and the same Principle, acting variously in various Circumstances?

Qu. 5. Do not Bodies and Light act mutually upon one another; that is to say, Bodies upon Light in emitting, reflecting, refracting and inflecting it, and Light upon Bodies for heating them, and putting their parts into a vibrating motion wherein heat consists?

Qu. 6. Do not black Bodies conceive heat more easily from Light than those of other Colours do, by reason that the Light falling on them is not reflected outwards, but enters the Bodies, and is often reflected and refracted within them, until it be stifled and lost?

DOI: 10.4324/9781003263814-3

Qu. 7. Is not the strength and vigor of the action between Light and sulphureous Bodies observed above, one reason why sulphureous Bodies take fire more readily, and burn more vehemently than other Bodies do?

[. . .]

Quest. 31. Have not the small Particles of Bodies certain Powers, Virtues, or Forces, by which they act at a distance, not only upon the Rays of Light for reflecting, refracting, and inflecting them, but also upon one another for producing a great Part of the Phænomena of Nature? For it's well known, that Bodies act one upon another by the Attractions of Gravity, Magnetism, and Electricity; and these Instances shew the Tenor and Course of Nature, and make it not improbable but that there may be more attractive Powers than these. For Nature is very consonant and conformable to her self. How these Attractions may be perform'd, I do not here consider. What I call Attraction may be perform'd by impulse, or by some other means unknown to me. I use that Word here to signify only in general any Force by which Bodies tend towards one another, whatsoever be the Cause. For we must learn from the Phænomena of Nature what Bodies attract one another, and what are the Laws and Properties of the Attraction, before we enquire the Cause by which the Attraction is perform'd. The Attractions of Gravity, Magnetism, and Electricity, reach to very sensible distances, and so have been observed by vulgar Eyes, and there may be others which reach to so small distances as hitherto escape Observation; and perhaps electrical Attraction may reach to such small distances, even without being excited by Friction.

[. . .]

Now the smallest Particles of Matter may cohere by the strongest Attractions, and compose bigger Particles of weaker Virtue; and many of these may cohere and compose bigger Particles whose Virtue is still weaker, and so on for divers Successions, until the Progression end in the biggest Particles on which the Operations in Chymistry, and the Colours of natural Bodies depend, and which by cohering compose Bodies of a sensible Magnitude. If the Body is compact, and bends or yields inward to Pression without any sliding of its Parts, it is hard and elastick, returning to its Figure with a Force rising from the mutual Attraction of its Parts. If the Parts slide upon one another, the Body is malleable or soft. If they slip easily, and are of a fit Size to be agitated by Heat, and the Heat is big enough to keep them in Agitation, the Body is fluid; and if it be apt to stick to things, it is humid; and the Drops of every fluid affect a round Figure by the mutual Attraction of their Parts, as the Globe of the Earth and Sea affects a round Figure by the mutual Attraction of its Parts by Gravity.

Since Metals dissolved in Acids attract but a small quantity of the Acid, their attractive Force can reach but to a small distance from them. And as in Algebra, where affirmative Quantities vanish and cease, there negative ones begin; so in Mechanicks, where Attraction ceases, there a repulsive Virtue ought to succeed. And that there is such a Virtue, seems to follow from the Reflexions and Inflexions of the Rays of Light. For the Rays are repelled by Bodies in both these Cases, without the immediate Contact of the reflecting or inflecting Body. It seems also to follow from the Emission of Light; the Ray so soon as it is shaken off from a shining Body by the vibrating Motion of the Parts of the Body, and gets beyond the reach of Attraction, being driven away with exceeding great Velocity. For that Force which is sufficient to turn it back in Reflexion, may be sufficient to emit it. It seems also to follow from the Production of Air and Vapour. The Particles when they are shaken off from Bodies by Heat or Fermentation, so soon as they are beyond the reach of the Attraction of the Body, receding from it, and also from one another with great Strength, and keeping at a distance, so as sometimes to take up above a Million of Times more space than they did before in the form of a dense Body. Which vast Contraction and Expansion seems unintelligible, by feigning the Particles of Air to be springy and ramous, or rolled up like Hoops, or by any other means than a repulsive Power. The Particles of Fluids which do not cohere too strongly, and are of such a Smallness as renders them most susceptible of those Agitations which keep Liquors in a Fluor, are most easily separated and rarified into Vapour, and in the Language of the Chymists, they are volatile, rarifying with an easy Heat, and condensing with Cold. But those which are grosser, and so less susceptible of Agitation, or cohere by a stronger Attraction, are not separated without a stronger Heat, or perhaps not without Fermentation. And these last are the Bodies which Chymists call fix'd, and being rarified by Fermentation, become true permanent Air; those Particles receding from one another with the greatest Force, and being most difficultly brought together, which upon Contact cohere most strongly. And because the Particles of permanent Air are grosser, and arise from denser Substances than those of Vapours, thence it is that true Air is more ponderous than Vapour, and that a moist Atmosphere is lighter than a dry one, quantity for quantity. From the same repelling Power it seems to be that Flies walk upon the Water without wetting their Feet; and that the Object-glasses of long Telescopes lie upon one another without touching; and that dry Powders are difficultly made to touch one another so as to stick together, unless by melting them, or wetting them with Water, which by exhaling may bring them together; and that two polish'd Marbles, which by immediate Contact stick together, are difficultly brought so close together as to stick.

And thus Nature will be very conformable to her self and very simple, performing all the great Motions of the heavenly Bodies by the Attraction of Gravity which intercedes those Bodies, and almost all the small ones of their Particles by some other attractive and repelling Powers which intercede the Particles. The *Vis inertiæ* is a passive Principle by which Bodies persist in their Motion or Rest, receive Motion in proportion to the Force impressing it, and resist as much as they are resisted. By this Principle alone there never could have been any Motion in the World. Some other

Principle was necessary for putting Bodies into Motion; and now they are in Motion, some other Principle is necessary for conserving the Motion. For from the various Composition of two Motions, 'tis very certain that there is not always the same quantity of Motion in the World. For if two Globes joined by a slender Rod, revolve about their common Center of Gravity with an uniform Motion, while that Center moves on uniformly in a right Line drawn in the Plane of their circular Motion; the Sum of the Motions of the two Globes, as often as the Globes are in the right Line described by their common Center of Gravity, will be bigger than the Sum of their Motions, when they are in a Line perpendicular to that right Line. By this Instance it appears that Motion may be got or lost. But by reason of the Tenacity of Fluids, and Attrition of their Parts, and the Weakness of Elasticity in Solids, Motion is much more apt to be lost than got, and is always upon the Decay. For Bodies which are either absolutely hard, or so soft as to be void of Elasticity, will not rebound from one another. Impenetrability makes them only stop. If two equal Bodies meet directly *in vacuo*, they will by the Laws of Motion stop where they meet, and lose all their Motion, and remain in rest, unless they be elastick, and receive new Motion from their Spring. If they have so much Elasticity as suffices to make them re-bound with a quarter, or half, or three quarters of the Force with which they come together, they will lose three quarters, or half, or a quarter of their Motion. And this may be try'd, by letting two equal Pendulums fall against one another from equal heights. If the Pendulums be of Lead or soft Clay, they will lose all or almost all their Motions: If of elastick Bodies they will lose all but what they recover from their Elasticity. If it be said, that they can lose no Motion but what they communicate to other Bodies, the consequence is, that *in vacuo* they can lose no Motion, but when they meet they must go on and penetrate one another's Dimensions. If three equal round Vessels be filled, the one with Water, the other with Oil, the third with molten Pitch, and the Liquors be stirred about alike to give them a vortical Motion; the Pitch by its Tenacity will lose its Motion quickly, the Oil being less tenacious will keep it longer, and the Water being less tenacious will keep it longest, but yet will lose it in a short time. Whence it is easy to understand, that if many contiguous Vortices of molten Pitch were each of them as large as those which some suppose to revolve about the Sun and fix'd Stars, yet these and all their Parts would, by their Tenacity and Stiffness, communicate their Motion to one another till they all rested among themselves. Vortices of Oil or Water, or some fluider Matter, might continue longer in Motion; but unless the Matter were void of all Tenacity and Attrition of Parts, and Communication of Motion, (which is not to be supposed,) the Motion would constantly decay. Seeing therefore the variety of Motion which we find in the World is always decreasing, there is a necessity of conserving and recruiting it by active Principles, such as are the cause of Gravity, by which Planets and

Comets keep their Motions in their Orbs, and Bodies acquire great Motion in falling; and the cause of Fermentation, by which the Heart and Blood of Animals are kept in perpetual Motion and Heat; the inward Parts of the Earth are constantly warm'd, and in some places grow very hot; Bodies burn and shine, Mountains take fire, the Caverns of the Earth are blown up, and the Sun continues violently hot and lucid, and warms all things by his Light. For we meet with very little Motion in the World, besides what is owing to these active Principles. And if it were not for these Principles, the Bodies of the Earth, Planets, Comets, Sun, and all things in them, would grow cold and freeze, and become inactive Masses; and all Putrefaction, Generation, Vegetation and Life would cease, and the Planets and Comets would not remain in their Orbs.

All these things being consider'd, it seems probable to me, that God in the Beginning form'd Matter in solid, massy, hard, impenetrable, moveable Particles, of such Sizes and Figures, and with such other Properties, and in such Proportion to Space, as most conduced to the End for which he form'd them; and that these primitive Particles being Solids, are incomparably harder than any porous Bodies compounded of them; even so very hard, as never to wear or break in pieces; no ordinary Power being able to divide what God himself made one in the first Creation. While the Particles continue entire, they may compose Bodies of one and the same Nature and Texture in all Ages: But should they wear away, or break in pieces, the Nature of Things depending on them, would be changed. Water and Earth, composed of old worn Particles and Fragments of Particles, would not be of the same Nature and Texture now, with Water and Earth composed of entire Particles in the Beginning. And therefore, that Nature may be lasting, the Changes of corporeal Things are to be placed only in the various Separations and new Associations and Motions of these permanent Particles; compound Bodies being apt to break, not in the midst of solid Particles, but where those Particles are laid together, and only touch in a few Points.

It seems to me farther, that these Particles have not only a *Vis inertiæ*, accompanied with such passive Laws of Motion as naturally result from that Force, but also that they are moved by certain active Principles, such as is that of Gravity, and that which causes Fermentation, and the Cohesion of Bodies. These Principles I consider, not as occult Qualities, supposed to result from the specifick Forms of Things, but as general Laws of Nature, by which the Things themselves are form'd; their Truth appearing to us by Phænomena, though their Causes be not yet discover'd. For these are manifest Qualities, and their Causes only are occult. And the *Aristotelians* gave the Name of occult Qualities, not to manifest Qualities, but to such Qualities only as they supposed to lie hid in Bodies, and to be the unknown Causes of manifest Effects: Such as would be the Causes of Gravity, and of magnetick and electrick Attractions, and of Fermentations, if we should suppose that these Forces or Actions arose from Qualities unknown to us, and uncapable of being discovered and made manifest. Such occult Qualities put a stop to the Improvement of natural Philosophy, and therefore of late Years have been rejected. To tell us that every Species of Things is endow'd with an occult specifick Quality by which it acts and produces manifest Effects, is to tell us nothing: But to derive two or three general Principles of Motion from Phænomena, and afterwards to tell us how the Properties and Actions of all corporeal Things follow from those manifest Principles, would be a very great step in Philosophy, though the Causes of those Principles were not yet discover'd: And therefore I scruple not to propose the Principles of Motion above-mention'd, they being of very general Extent, and leave their Causes to be found out.

Now by the help of these Principles, all material Things seem to have been composed of the hard and solid Particles above-mention'd, variously associated in the first Creation by the Counsel of an intelligent Agent. For it became him who created them to set them in order. And if he did so, it's unphilosophical to seek for any other Origin of the World, or to pretend that it might arise out of a Chaos by the mere Laws of Nature; though being once form'd, it may continue by those Laws for many Ages. For while Comets move in very excentrick Orbs in all manner of Positions, blind Fate could never make all the Planets move one and the same way in Orbs concentrick, some inconsiderable Irregularities excepted, which may have risen from the mutual Actions of Comets and Planets upon one another, and which will be apt to increase, till this System wants a Reformation. Such a wonderful Uniformity in the Planetary System must be allowed the Effect of Choice. And so must the Uniformity in the Bodies of Animals, they having generally a right and a left side shaped alike, and on either side of their Bodies two Legs behind, and either two Arms, or two Legs, or two Wings before upon their Shoulders, and between their Shoulders a Neck running down into a Back-bone, and a Head upon it; and in the Head two Ears, two Eyes, a Nose, a Mouth, and a Tongue, alike situated. Also the first Contrivance of those very artificial Parts of Animals, the Eyes, Ears, Brain, Muscles, Heart, Lungs, Midriff, Glands, Larynx, Hands, Wings, swimming Bladders, natural Spectacles, and other Organs of Sense and Motion; and the Instinct of Brutes and Insects, can be the effect of nothing else than the Wisdom and Skill of a powerful ever-living Agent, who being in all Places, is more able by his Will to move the Bodies within his boundless uniform Sensorium, and thereby to form and reform the Parts of the Universe, than we are by our Will to move the Parts of our own Bodies. And yet we are not to consider the World as the Body of God, or the several Parts thereof, as the Parts of God. He is an uniform Being, void of Organs, Members or Parts, and they are his Creatures subordinate to him, and subservient to his Will; and he is no more the Soul of them, than the Soul of Man is the Soul of the Species of Things carried through the Organs of Sense into the place of its Sensation, where it perceives them by means of its immediate Presence, without the Intervention of any third thing. The Organs of Sense are not for enabling the Soul to perceive the Species of Things in its Sensorium, but only for conveying them thither; and God has no need of such Organs, he being every where present to the Things themselves. And since Space is divisible *in infinitum*, and Matter is not necessarily in all places, it may be also allow'd that God is able to create Particles of Matter of several Sizes and Figures, and in several Proportions to Space, and perhaps of different Densities and Forces, and thereby to vary the Laws of Nature, and make Worlds of several sorts in several Parts of the Universe. At least, I see nothing of Contradiction in all this.

As in Mathematicks, so in Natural Philosophy, the Investigation of difficult Things by the Method of Analysis, ought ever to precede the Method of Composition. This Analysis consists in making Experiments and Observations, and in drawing general Conclusions from them by Induction, and admitting of no Objections against the Conclusions, but such as are taken from Experiments, or other certain Truths. For Hypotheses are not to be regarded in experimental Philosophy. And although the arguing from Experiments and Observations by Induction be no Demonstration of general Conclusions; yet it is the best way of arguing which the Nature of Things admits of, and may be looked upon as so much the stronger, by how much the Induction is more general. And if no Exception occur from Phænomena, the Conclusion may be pronounced generally. But if at any time afterwards any Exception shall occur from Experiments, it may then begin to be pronounced with such Exceptions as occur. By this way of Analysis we may proceed from Compounds to Ingredients, and from Motions to the Forces producing them; and in general, from Effects to their Causes, and from particular Causes to more general ones, till the Argument end in the most general. This is the Method of Analysis: And the Synthesis consists in assuming the Causes discover'd, and establish'd as Principles, and by them explaining the Phænomena proceeding from them, and proving the Explanations.

In the two first Books of these Opticks, I proceeded by this Analysis to discover and prove the original Differences of the Rays of Light in respect of Refrangibility, Reflexibility, and Colour, and their alternate Fits of easy Reflexion and easy Transmission, and the Properties of Bodies, both opake and pellucid, on which their Reflexions and Colours depend. And these Discoveries being proved, may be assumed in the Method of Composition for explaining the Phænomena arising from them: An Instance of which Method I gave in the End of the first Book. In this third Book I have only begun the Analysis of what remains to be discover'd about Light and its Effects upon the Frame of Nature, hinting several things about it, and leaving the Hints to be examin'd and improv'd by the farther Experiments and Observations of such as are inquisitive. And if natural Philosophy in all its Parts, by pursuing this Method, shall at length be perfected, the Bounds of Moral Philosophy will be also enlarged. For so far as we can know by natural Philosophy what is the first Cause, what Power he has over us, and what Benefits we receive from him, so far our Duty towards him, as well as that towards one another, will appear to us by the Light of Nature. And no doubt, if the Worship of false Gods had not blinded the Heathen, their moral Philosophy would have gone farther than to the four Cardinal Virtues; and instead of teaching the Transmigration of Souls, and to worship the Sun and Moon, and dead Heroes, they would have taught us to worship our true Author and Benefactor, as their Ancestors did under the Government of *Noah* and his Sons before they corrupted themselves.

Editorial Headnote

3. Émilie du Châtelet, "Of Hypotheses," from *Foundations of Physics* [*Institutions de physique*], tr. Isabelle Bour and Judith P. Zinsser (1740, tr. 2009)

Concern about Newton's uncertain relationship with hypotheses was not merely a product of the nineteenth century – it was a theme highlighted as important even by Newton's first commentators across the hundred years prior. Of these, Émilie du Châtelet (1706–1749) is arguably the most influential. Her translation of Newton's *Principia* into French, along with her detailed commentary and her own synthetic works on metaphysics and physics, from one of which this excerpt is drawn, were largely responsible (along with, or perhaps better, via, her romantic companion Voltaire) for introducing Newtonian ideas to the Francophone world and to continental Europe more generally.

Du Châtelet's presentation of the dilemma which hypotheses pose for scientific methodology is as canonical as one might imagine. For all that we seem to have a genuine need for hypothesis – after all, science must begin somewhere – scientific works have tended to fall into one of two extremes regarding their use. To those taking a more hard-line, "anti-hypothesis" reading of Newton, she argues that Newtonians have been too quick to think that they can reason entirely without hypotheses. Cartesians, on the other hand, have been too quick to invent hypothetical mechanisms in the absence of evidence. Du Châtelet thus frames precisely the problem that will motivate much general work on methodology in the years to come: what standards ought we have in order to ensure that we are examining useful, worthwhile hypotheses rather than errant flights of fancy?

She goes on to offer us among the first sketches for a theory of such standards ever developed. Any valid hypothesis must be philosophically coherent, must be drawn from complete data, and must not be given undue epistemic credit (recall the discussion of Newton's "feigning" in the headnote to the previous text). She also carefully weighs the role of experiment in the testing of hypotheses, in ways that would be inspirational for a number of the later authors we will consider in the rest of this volume. First, she approvingly cites the power of an experimental test of a hypothesis, such that the hypothesis "not only explains the phenomenon that one had proposed to explain with it, but also that all the consequences drawn from it agree with the observations." She also explores the role of experiments in falsifying, rather than confirming, hypotheses, either in whole or in part, as well as offering an injunction against *ad hoc* formulation of modifications to our hypotheses.

Finally, in addition to a clear presentation of the dilemma surrounding the use of hypotheses and steps toward its resolution, du Châtelet further elucidates the stakes of the argument, canvassing just how important hypothetical reasoning has in fact been to crucial works throughout the recent history of science. While she does not systematize her approach to scientific methodology to the extent that a later figure like Herschel or Whewell will, she argues forcefully both for its importance and for a variety of the most important aspects that frame the question throughout the decades to come.

3

OF HYPOTHESIS FROM *FOUNDATIONS OF PHYSICS*

Émilie Du Châtelet

Source: Émilie Du Châtelet, "Of Hypothesis," from *Foundations of Physics*, tr. Isabelle Bour and Judith P. Zinsser (2009 [1740]), pp. 147–155.

CHAPTER FOUR

Of hypotheses

THE USEFULNESS OF PROBABILITIES IN PHYSICS.

§.53. The true causes of natural effects and of the phenomena we observe are often so far from the principles on which we can rely and the experiments we can make that one is obliged to be content with probable reasons to explain them. Thus, probabilities are not to be rejected in the sciences, not only because they are often of great practical use, but also because they clear the path that leads to the truth.

USE OF HYPOTHESIS.

§.54. There must be a beginning in all researches, and this beginning must almost always be a very imperfect, often unsuccessful attempt. There are unknown truths just as there are unknown countries to which one can only find the good route after having tried all the others. Thus, some must run the risk of losing their way in order to mark the good path for others; so it would be doing the sciences great injury, infinitely delaying their progress, to banish hypotheses as some modern philosophers have.

MISUSES OF HYPOTHESES BY THE DISCIPLES OF DESCARTES.

§.55. Descartes, who had established much of his philosophy on hypotheses, because it was almost impossible to do otherwise in his time, gave the whole learned world a taste for hypotheses; and it was not long before one fell into a taste for fictions. Thus, the books of philosophy, which should have been collections of truths, were filled with fables and reveries.

DOI: 10.4324/9781003263814-4

THE DISCIPLES OF M. NEWTON HAVE FALLEN INTO THE OPPOSITE EXCESS.

M. Newton, and above all his disciples, have fallen into the opposite excess: disgusted with suppositions and errors that they found filled books of philosophy, they rose up against hypotheses and tried to make them suspect and ridiculous, by calling them *the poison of reason and the plague of Philosophy.* Moreover, he alone, who was able to assign and demonstrate the causes of all that we see, would be entitled to banish hypotheses from physics; but, as for us, who do not seem to be cut out for such knowledge, and who can only arrive at the truth by crawling from probability to probability, it is not for us to pronounce so boldly against hypotheses.

HOW HYPOTHESES ARE MADE.

§.56. When certain things are used to explain what has been observed, and though the truth of what has been supposed is impossible to demonstrate, one is making a hypothesis. Thus, philosophers frame hypotheses to explain the phenomena, the cause of which cannot be discovered either by experiment or by demonstration.

HYPOTHESES ARE THE THREADS THAT LEAD US TO THE MOST SUBLIME DISCOVERIES.

§.57. If we take the trouble to study the way in which the most sublime discoveries were made, we will see that success only came after many unnecessary hypotheses had been made and yet the duration and unprofitableness of this work had not proved discouraging; for hypotheses are often the only available means to discover new truths. It is true that the means is slow and requires labor all the more onerous because for a long time one cannot know if it will be useful or fruitless. Similarly, when one takes an unknown route and finds several paths, it is only after walking a long time that one can be sure if one has taken the good route or if one has been mistaken. But if the uncertainty over which of these routes is the right one were a reason not to take any of them, it is certain that one would never arrive; whereas when one has the courage to set off, there is no doubt that of the three routes, of which two have misled us, the third will infallibly lead us to our goal.

WITHOUT HYPOTHESIS FEW DISCOVERIES WOULD HAVE BEEN MADE IN ASTRONOMY.

It is in this manner that astronomy has brought us to the point where we admire it today; for, if, to calculate the path of the celestial bodies, astronomers had waited until the true theory of the planets had been found, there would be no astronomy now.

IT IS TO THESE THAT WE OWE THE TRUE SYSTEM OF THE WORLD.

The first idea of those who applied themselves to this science, just like the first idea of all other men, must have been that the Sun and all the celestial bodies turned around the Earth in twenty-four hours. Thus, they began to explain and to predict phenomena by this hypothesis, called *Ptolemy's hypothesis,* until the insurmountable difficulties of the consequences that derived from it when compared with observations, and the impossibility of constructing tables according to this hypothesis which were in accord with the phenomena of the sky, brought Copernicus to abandon it entirely and to test the opposite hypothesis, which is so much in agreement with the phenomena, that its certitude is at present not far from demonstration; and that no astronomer dares adopt that of Ptolemy.

THEY OFTEN GIVE THE IDEA OF HOW TO DO NEW, VERY USEFUL EXPERIMENTS.

§.58. Hypotheses must then find a place in the sciences, since they promote the discovery of truth and offer new perspectives; for when a hypothesis is once posed, experiments are often done to ascertain if it is a good one, experiments which would never have been thought of without it. If it is found that these experiments confirm it, and that it not only explains the phenomenon that one had proposed to explain with it, but also that all the consequences drawn from it agree with the observations, its probability grows to such a point that we cannot refuse our assent to it, and that is almost equivalent to a demonstration.

The example of astronomers can further serve marvelously well to clarify this matter; for the true orbits of the planets were ascertained by first supposing they made their revolutions in circles, of which the Sun occupied the center; but the variation in their speed and their apparent diameters being contradictory to this hypothesis, it was supposed they moved in eccentric circles, that is to say, in circles of which the Sun did not occupy the center. This supposition, which corresponded to the movements of the Earth well enough, deviated greatly from what is observed about the planet Mars; and to remedy this, attempts were made to make a new correction to the curve the planets describe in their annual revolution. This procedure succeeded so well that finally Kepler, going from supposition to supposition, found their true orbit, which admirably corresponded to all the appearances of the planets, and this orbit is an ellipse, of which the Sun occupies one of the foci.

By means of this hypothesis of the elliptic nature of orbits Kepler came to discover the proportionality of the areas and the times, and that of the times and the distances. And these two famous theorems, called the *Analogies of Kepler,* put M. Newton in reach of demonstrating that the supposition of the elliptic nature of the planets' orbits agrees with the laws of mechanics, and of assigning the proportion of forces that direct the movements of celestial bodies.

Thus, it is evident that it is to hypotheses first made and then corrected that we are indebted for the beautiful and sublime knowledge of which astronomy and its subsidiary sciences are filled at present. It is impossible to see how men could have arrived there by other means.

IT IS BY MEANS OF HYPOTHESES THAT M. HUYGENS DISCOVERED THAT SATURN WAS SURROUNDED BY A RING.

By the same means we know today that Saturn is surrounded by a ring that reflects light, and is separated from the body of the planet, and inclined to the ecliptic. For M. Huygens, who discovered it first, did not observe what astronomers now describe; but he observed several phases of it, which sometimes resembled nothing less than a ring. Next, comparing the successive changes of these phases and all the observations that he had made of it, he sought a hypothesis that fit them, and explained these different appearances. That of a ring succeeds very well; it not only explains the appearances but also predicts the phases of this ring very accurately.

This correspondence between hypothesis and observation has finally converted this supposition of M. Huygens into a certainty; and no one doubts now that this ring is very real; thus, hypotheses brought us this beautiful discovery of the ring of Saturn.

The same can be said of the ingenious explanation the same M. Huygens gave for halos, that is to say, of these sorts of colored crowns that sometimes appear around stars. No one before him had conceived of what might be the cause of these phenomena; but M. Huygens, after several fruitless suppositions, finally found that, by supposing the air to be filled with frozen grains of hail with a kernel of snow in their center, all the circumstances that accompany these phenomena could be explained, and no one has ventured to call into question M. Huygens's explanation.

DIVISION IS FOUNDED ON HYPOTHESES ONLY.

§.59. It is the same with numbers. Division, for example, is founded on hypotheses only. Without hypotheses you could not divide, for when you begin division, you suppose that the divisor is contained in the dividend as many times as the first number of the divisor is contained in the first number, or in the first two numbers of the dividend; and then you verify this supposition by multiplying the divisor by the quotient, and by subtracting from the dividend the product of this multiplication. If you find that this subtraction cannot be done, you conclude that the quotient is too big, and then you correct it. This whole operation is done by means of hypotheses.

HYPOTHESES ARE NOT ONLY VERY USEFUL, BUT EVEN SOMETIMES VERY NECESSARY.

§.60. So making hypotheses is allowed, and it is even very useful in all cases when we cannot discover the true reason for a phenomenon and the attendant circumstances, neither a priori, by means of truths that we already know; nor a posteriori, with the help of experiments.

HOW ONE MUST PROCEED WHEN MAKING A HYPOTHESIS.

§.61. Without doubt there are rules to follow and pitfalls to avoid in hypotheses. The first is, that it not be in contradiction with the principle of sufficient reason, nor with any principles that are the foundations of our knowledge. The second rule is to have certain knowledge of the facts that are within our reach, and to know all the circumstances attendant upon the phenomena we want to explain. This care must precede any hypothesis invented to explain it; for he who would hazard a hypothesis without this precaution would run the risk of seeing his explanation overthrown by new facts that he had neglected to find out about. This is what happened to him who wanted to explain electricity after having only seen how Spanish wax, rubbed vigorously, attracts bits of paper. For it would have been easy to make other bodies do what happens with Spanish wax; rubbed in the same way, they would also have been electrified. Thus, the explanation of electricity by Spanish wax alone had been insufficient and precipitous.

When one can hope to know the greatest number of circumstances attendant upon a phenomenon, then one can seek the reason for it by means of hypotheses, taking the risk of having to correct it or having to be corrected; but the efforts made to find the truth are always glorious, even though they might be fruitless.

PITFALL TO AVOID IN HYPOTHESES.

§.62. Since hypotheses are only made in order to discover the truth, they must not be passed off as the truth itself, before one is able to give irrefutable proofs. So it is very important for the progress of the sciences not to delude oneself and others with the hypotheses one has invented, but one should estimate the degree of probability in a hypothesis, and never impose it on others by detours and a semblance of demonstration, which has much too often led people with a thirst for knowledge into error.

With this precaution one does not run the danger of taking for certain that which is not; and one inspires those who follow us to correct the faults in our hypotheses and to provide what they lack to make them certain.

§.63. Most of those who, since Descartes, have filled their writings with hypotheses to explain facts, which very often they only knew imperfectly, have sinned against this rule and have tried to pass off their suppositions as truths; this is partly the source of the disgust for hypotheses in this century. But the excessive resort to a useful thing does not detract from its utility and must not prevent us from making use of it when this can bear fruit.

A SINGLE CONTRARY EXPERIMENT SUFFICES TO REJECT A HYPOTHESIS.

§.64. One experiment is not enough for a hypothesis to be accepted, but a single one suffices to reject it when it is contrary to it. It follows, for example, from the hypothesis in which one supposes that the Sun moves around the Earth, which serves as the center of its orbit, that the diameters of the Sun must be equal at all times of the year; but experience shows that they appear unequal. From this observation, one can therefore conclude with certainty that the hypothesis, of which this equality is a consequence is false, and that the Earth does not at all occupy the center of the Sun's orbit.

A HYPOTHESIS CAN BE TRUE IN ONE OF ITS PARTS AND FALSE IN ANOTHER.

§.65. A hypothesis may be true in one of its parts and false in another; then the part that is found to be in contradiction with experiment must be corrected.

But great care must be taken to put in the conclusion only what must be there, and not to attribute to the entire hypothesis a fault which only applies to one of its parts. For example, M. Descartes attributed the fall of bodies toward the center of the Earth to a vortex of fluid matter impelling bodies to move toward the center by its rapid swirling around the Earth; but M. Huygens demonstrated by an irrefutable experiment that, according to this supposition, bodies should be directed in a fall perpendicular to the axis of the Earth and not toward its center. It can be concluded from that, that a vortex of fluid matter, such as M. Descartes conceived, would not cause bodies to fall toward the center of the Earth; but it would be too precipitous to conclude that no fluid material caused the phenomenon of the fall of bodies. It is the same with the other vortices, which, according to M. Descartes, carry the planets around the Sun; for M. Newton demonstrated that this supposition did not agree with the laws of Kepler. So it must be inferred that the movements of the planets are not the effect of the vortices of fluid material that M. Descartes supposed explained them. But it cannot be legitimately concluded that a vortex, or several vortices, conceived of in a different way, cannot be the cause of these movements.

§.66. Thus, in making a hypothesis one must deduce all the consequences that can legitimately be deduced, and next compare them with experiment; for should all these consequences be confirmed by experiments, the probability would be greatest. But if there is a single one contrary to them, either the entire hypothesis must be rejected, if this consequence follows from the entire hypothesis, or that part of the hypothesis from which it necessarily follows.

Astronomers give us another example of this rule; indeed, a plethora of discoveries would not have come about in astronomy if no attempt had been made to verify by experiment what was deduced from hypotheses. It follows, for example, from Copernicus's hypothesis that if the distance of a star to the Earth has a relationship comparable to the diameter of its orbit, the height of the pole and the fixed stars must vary at different times of the year. The desire to verify this consequence led several astronomers to make observations on this annual parallax or height of the fixed stars; among others, M. Bradley, in whose hands this consequence was not only confirmed but then gave birth to this beautiful theory of the aberration of the fixed stars, which would never have been thought of before.

DEFINITION OF HYPOTHESES. WHAT MAKES THEM PROBABLE.

§.67. Hypotheses, then, are only probable propositions that have a greater or lesser degree of certainty, depending on whether they satisfy a more or less great number of circumstances attendant upon the phenomenon that one wants to explain by their means. And, as a very great degree of probability gains our assent, and has on us almost the same effect as certainty, hypotheses finally become truths when their probability increases to such a point that one can morally present them as a certainty; this is what happened with Copernicus's system of the world, and with M. Huygens's on the ring of Saturn.

WHAT MAKES THEM INVALID.

By contrast, a hypothesis becomes all the more improbable as it fails to explain more of its attendant circumstances, as in Ptolemy's hypothesis.

§.68. When a hypothesis is made, one must have reasons for preferring the supposition on which it is founded to all other suppositions; otherwise one spews forth chimeras, and precarious principles that have no foundation.

§.69. So it is necessary not only that all one supposes be possible, but also that it be possible in the manner one uses it; and that the phenomena result necessarily, and without the obligation to make new suppositions. Otherwise, the supposition does not deserve the name of hypothesis; for a hypothesis is a supposition that explains a phenomenon. When the necessary consequences do not follow from it, and to explain the phenomenon, a new hypothesis must be created in order to use the first, this hypothesis is only a fiction unworthy of a philosopher.

§.70. If those who wanted to explain so many surprising effects by means of hooked, branchlike, and serrated particles had paid attention to what is required to make a truly philosophical hypothesis, they would not have slowed, as they did, the progress of the sciences by creating monsters that subsequently had to be fought against as realities.

HYPOTHESES ARE ONE OF THE GREAT MEANS OF THE ART OF INVENTION. GOOD HYPOTHESES HAVE ALWAYS BEEN MADE BY THE GREATEST MEN.

§.71. By distinguishing between the good and the bad use of hypotheses, both extremes are avoided, and without giving oneself up to fictions, a method very necessary to the art of invention is not denied to the sciences, a method that is the only means that can be used in difficult researches requiring correction over several centuries and the work of several men before attaining a certain perfection. And it must not be feared that by this method philosophy might become a heap of fables; for we have seen that a good hypothesis can only be made when a great number of facts and circumstances attendant upon the phenomenon one wants to explain have been observed (§.61), and that the hypothesis is only true and only deserves to be adopted when it explains all the circumstances (§.66). Therefore, the good hypotheses will always be the work of the greatest men. Copernicus, Kepler, Huygens, Descartes, Leibniz, M. Newton himself, have all imagined useful hypotheses to explain complicated and difficult phenomena; and the examples of these great men and their success must show how much those who want to banish hypotheses from philosophy misunderstand the interests of the sciences.

Editorial Headnote

4. Immanuel Kant, "Preface," from *Metaphysical Foundations of Natural Science*, tr. Ernest Belfort Bax (1786, tr. 1883)

As has increasingly come to be appreciated in the philosophical community, the works of Immanuel Kant (1724–1804) owe a substantial debt to the physics of Newton. As Kant's œuvre constitutes one of the most important systematic philosophical efforts of the late eighteenth century, it is also an important factor in any presentation of nineteenth-century thought about scientific methodology – especially, as we will see, for the works of William Whewell.

Kant's goal here is ambitious – no less than to derive, by reflecting upon motion and bodies as accessible to us in our experience, the laws of Newtonian mechanics and the nature of its basic forces, along with a conception of absolute space that is freed from worries about its epistemic inaccessibility. As in Kant's other writings, such as the *Critique of Pure Reason* (1781), this will not be a purely *a priori* argument, but a transcendental one – a presentation of absolute space as a kind of "regulative ideal," a guide for our reasoning, but one which our reasoning can never quite fully comprehend.

As Michael Friedman notes, the work "is difficult to read, even by Kantian standards."[1] This excerpt constitutes the work's preface, where Kant is occupied with clarifying what he means by 'science.' By contrast with Newton, Kant argues that if "the grounds or principles [a science] contains are in the last resort merely empirical, as, for instance, in chemistry," then "the whole does not in strictness deserve the name of science." On the contrary, a true science is one for which "the natural laws at its foundation are cognized *a priori*, and are not mere laws of experience" (p. 34). We thus are not looking merely for well-confirmed laws, much less for hypotheses, but for principles which can effectively guarantee the certainty of the conclusions drawn from them – a much higher bar than that recognized in the rest of the Newtonian tradition.

How are we to arrive at *a priori* knowledge in scientific maters, knowledge which Kant will call the "*pure* portion" (p. 34) of such a science? First and foremost, Kant famously claims that "a doctrine of nature can only contain so much science proper as there is in it of applied mathematics" (p. 35) – for, Kant argues, it is only by building mathematical constructions that we can obtain a fully general, *a priori* proof of the possibility of a concept. (He despairs, in passing, for the future of both chemistry and psychology.) We must therefore take our concepts of matter, space, and time, and attempt to construct as much as we can from them in as much generality as possible, in order to arrive at what Kant calls metaphysical principles.

Kant divides these principles into four parts, which are the subject of the remainder of the book.[2] First, in what he calls the *phoronomy*, we are to imagine motion only as a *quantity*. Put differently, how is it that motion can be rendered the object of mathematical quantification in the first place? To do so, we need to know how velocities (possibly velocities with respect to different reference frames) can be added together. Second, in the *dynamics*, we add to this motion the fact that bodies *fill a space*. To do so, Kant claims, we need to understand how the forces of attraction and repulsion of their parts work together to arrive at stability, introducing (of necessity) the idea of Newtonian forces. Third, in the *mechanics*, we add the phenomena of the transfer of motion from one body to another (e.g., via impact), leading us to the concept of Newtonian momentum as well as derivations of the conservation of mass, the law of inertia, and Newton's third law (i.e., that every action leads to an equal and opposite reaction). Fourth and finally, in the *phenomenology*, Kant returns to the question of space – in particular, how we can say absolutely that a particular body is at rest or in motion, despite the fact that we constantly measure velocities with respect to different reference frames. We do so, roughly, by inferring our way toward the common center of mass of the universe.

The details are much more complex than this simple presentation can properly describe, and are still the matter of extensive discussion and debate among Kant scholars. For our purposes here, what counts is the radically different understanding of proper scientific method that Kant's work embodies. He places an enormous emphasis on the role of *a priori* and mathematical reasoning in the construction of scientific concepts. How should the nature of the fundamental concepts upon which our sciences are based be related to the use of those concepts in scientific theorizing? What role should *a priori* or metaphysical reasoning play in empirical sciences? These questions would only become more pressing as the concepts involved in those sciences became immensely more complex over the course of the nineteenth century.

Notes

1 M. Friedman, 'Introduction,' in M. Friedman (ed.), *Immanuel Kant: Metaphysical Foundations of Natural Science* (Cambridge: Cambridge University Press, 2004), pp. vii–xxx, on p. xxviii.

2 For more detail, see Friedman, 'Introduction,' pp. xv–xxvii, which this presentation follows.

4

PREFACE TO *METAPHYSICAL FOUNDATIONS OF NATURAL SCIENCE*

Immanuel Kant

Source: Immanuel Kant, "Preface" from *Metaphysical Foundations of Natural Science*, from *Prolegomena and Metaphysical Foundations of Natural Science,* tr. Ernest Belfort Bax (1883 [1786]), pp. 137–149.

THE METAPHYSICAL FOUNDATIONS OF NATURAL SCIENCE

PREFACE

IF the word Nature be merely taken in its *formal* signification, there may be as many natural sciences as there are specifically different things (for each must contain the inner principle special to the determinations pertaining to its existence), inasmuch as it [Nature] signifies the primal inner principle of all that belongs to the existence of a thing.[1] But Nature, regarded in its *material* significance, means not a quality, but the sum-total of all things, in so far as they can be *objects of our senses,* and therefore of experience; in short, the totality of all phenomena – the sense-world, exclusive of all nonsensuous objects. Now Nature, in this sense of the word, has two main divisions, in accordance with the main distinction of our sensibility, one of which comprises the objects of the *outer,* the other the object of the *inner* sense; thus rendering possible a two-fold doctrine of Nature, the DOCTRINE OF BODY and the DOCTRINE OF SOUL, the first dealing with *extended,* and the second with *thinking,* Nature.

Every doctrine constituting a system, namely, a whole of cognition, is termed a science; and as its principles may be either axioms of the *empirical* or *rational* connection of cognitions in a whole, so natural science, whether it be doctrine of body or doctrine of soul, would have to be divided into *historical* and *rational* natural science, were it not that the word *nature* (as implying the deduction of the manifold pertaining to the existence of things, from its *inner principle*) necessitates a knowledge through reason of its system, if it is to deserve the name natural science. Hence, doctrine of nature may be better divided into *historical doctrine of nature,* comprising nothing but systematically-ordered facts respecting natural things (which again would consist of *description of nature* as a system of classes according to resemblances, and *history of nature* as a systematic presentation of

DOI: 10.4324/9781003263814-5

the same at different times and in different places), and *natural science.* Natural science, once more, would be either natural science *properly* or *improperly* so-called, of which the first would treat its subject wholly according to principles *à priori,* and the second according to laws derived from experience.

That only can be called science (*wissenschaft*) *proper* whose certainty is apodictic: cognition that can merely contain empirical certainty is only improperly called science. A whole of cognition which is systematic is for this reason called *science,* and, when the connection of cognition in this system is a system of causes and effects, *rational* science. But when the grounds or principles it contains are in the last resort merely empirical, as, for instance, in chemistry, and the laws from which the reason explains the given facts are merely empirical laws, they then carry no consciousness of their *necessity* with them (they are not apodictically certain), and thus the whole does not in strictness deserve the name of science; chemistry indeed should be rather termed systematic art than science.

A rational doctrine of nature deserves the name of natural science only when the natural laws at its foundation are cognised *à priori,* and are not mere laws of experience. A natural cognition of the first kind is called *pure,* that of the second *applied,* rational cognition. As the word nature itself carries with it the conception of law, and this again the conception of the *necessity* of all the determinations of a thing appertaining to its existence, it is easily seen why natural science must deduce the legitimacy of its designation only from a *pure* part of it, [a part] namely, which contains the principles *à priori* of all remaining natural explanations, and why only by virtue of this portion it is properly science, in such wise, that, according to the demands of the reason, all natural knowledge must at last turn on natural science and there find its conclusion. This is because the above necessity of law inseparably attaches to the conception of nature, and hence must be thoroughly comprehended. For this reason the most complete explanation of particular phenomena upon chemical principles, invariably leaves an unsatisfactoriness behind it, because from these accidental laws, learnt by mere experience, no grounds *à priori* can be adduced.

Thus all natural science *proper* requires a *pure* portion, upon which the apodictic certainty required of it by the reason can be based; and inasmuch as this is in its principles wholly heterogeneous from those which are merely empirical, it is at once a matter of the utmost importance, indeed in the nature of the case, as regards method of indispensable duty, to expound this part separately and unmixed with the other, and as far as possible in its completeness; in order that we may be able to determine precisely what the reason can accomplish for itself, and where its capacity begins to require the assistance of empirical principles. Pure cognition of the reason from mere *conceptions* is called pure philosophy or metaphysics, while that which only bases its cognition on the *construction* of conceptions, by means of the presentation of the object in an *à priori* intuition, is termed mathematics.

What may be called natural science proper presupposes metaphysics of nature; for laws, i.e. principles of the necessity of that which belongs to the *existence* of a thing, are occupied with a conception which does not admit of construction, because its existence cannot be presented in any *à priori* intuition; natural science proper, therefore, presupposes metaphysics. Now this must indeed always contain exclusively principles of a non-empirical origin (for, for this reason it bears the name of metaphysics); but it may be *either* without reference to any definite object of experience, and therefore undetermined as regards the nature of this or that thing of the sense-world, and treat of the laws rendering possible the conception of nature in general, in which case it is the *transcendental* portion of the metaphysics of nature; *or* it may occupy itself with the particular nature of this or that kind of thing, of which an empirical conception is given, in such wise, that except what lies in this conception, no other empirical principle will be required for its cognition. For instance: it lays the empirical conception of a matter, or of a thinking entity, at its foundation, and searches the range of the cognition of which the reason is *à priori* capable respecting these objects; and thus, though such a science must always be termed a metaphysic of nature (namely, of corporeal or thinking nature), it is then not a universal but a *particular* metaphysical natural science (physics and psychology), in which the above transcendental principles are applied to the two species of sense-objects. But I maintain that in every special natural doctrine only so much science *proper* is to be met with as mathematics; for, in accordance with the foregoing, science proper, especially [science] of nature, requires a pure portion, lying at the foundation of the empirical, and based upon an *à priori* knowledge of natural things. Now to cognise anything *à priori* is to cognise it from its mere possibility; but the possibility of determinate natural things cannot be known from mere conceptions; for from these the possibility of the thought (that it does not contradict itself) can indeed be known, but not of the object, as natural thing which can be given (as existent) outside the thought. Hence, to the possibility of a determinate natural thing, and therefore to cognise it *à priori,* is further requisite that the *intuition* corresponding *à priori* to the conception should be given; in other words, that the conception should be constructed. But cognition of the reason through construction of conceptions is mathematical. A pure philosophy of nature in general, namely, one that only investigates what constitutes a nature in general, may thus be possible without mathematics; but a pure doctrine of nature respecting *determinate* natural things (corporeal doctrine and mental doctrine), is only possible by means of mathematics; and as in every natural doctrine only so much science proper is to be met with therein as there is cognition *à priori,* a doctrine of nature can only contain so much science proper as there is in it of applied mathematics.

So long, therefore as no conception is discovered for the chemical effects of substances on one another, which admits of being constructed, that is, no law of the approach or retreat of the parts can be stated in accordance with which (as, for instance, in proportion to their densities) their motions, together with the consequences of these, can be intuited and presented *à priori* (a demand that will scarcely ever be fulfilled), chemistry will be nothing more than a systematic art or experimental doctrine, but never science proper, its principles being merely empirical and not admitting of any presentation *à priori*; as a consequence, the principles of chemical phenomena cannot make their possibility in the least degree conceivable, being incapable of the application of mathematics.

But still farther even than chemistry must empirical psychology be removed from the rank of what may be termed a natural science proper; firstly, because mathematics is inapplicable to the phenomena of the internal sense and its laws, unless indeed we consider merely the *law of permanence* in the flow of its internal changes; but this would be an extension of cognition, bearing much the same relation to that procured by the mathematics of corporeal knowledge, as the doctrine of the properties of the straight line does to the whole of geometry; for the pure internal intuition in which psychical phenomena are constructed is time, which has only one dimension. But not even as a systematic art of analysis, or experimental doctrine, can it ever approach chemistry, because in it the manifold of internal observation is only separated in thought, but cannot be kept separate and be connected again at pleasure; still less is another thinking subject amenable to investigations of this kind, and even the observation itself, alters and distorts the state of the object observed. It can never therefore be anything more than an historical, and as such, as far as possible systematic natural doctrine of the internal sense, i.e. a natural description of the soul, but not a science of the soul, nor even a psychological experimental doctrine. This is the reason why, in the title of this work, which, properly speaking, contains the axioms of corporeal doctrine, we have employed, in accordance with the usual custom, the general name of natural science, because this designation in the strict sense is applicable to it alone, and hence occasions no ambiguity.

But to render possible the application of mathematics to the doctrine of body, by which alone it can become natural science, principles of the *construction* of conceptions belonging to the possibility of matter in general must precede. Hence a complete analysis of the conception of a matter in general must be laid at its foundation; this is the business of pure philosophy, which for the purpose makes use of no special experiences, but only of those which it meets with in separate (although in themselves empirical) conceptions, with reference to pure intuitions in space and time (according to laws, essentially depending on the conception of nature in general), thus constituting it a real *metaphysic of corporeal nature.*

All natural philosophers, who wished to proceed mathematically in their work, have hence invariably (although unknown to themselves) made use of metaphysical principles, and must make use of such, it matters not how energetically they may otherwise repudiate any claim of metaphysics on their science. Without doubt by the latter they understood the illusion of manufacturing possibilities at pleasure, and playing with conceptions, perhaps quite incapable of being presented in intuition, and possessing no other guarantee of their objective reality than that they do not stand in contradiction with themselves. But all true metaphysics is taken from the essential nature of the thinking faculty itself, and therefore in nowise invented, since it is not borrowed from experience, but contains the pure operations of thought, that is, conceptions and principles *à priori,* which the manifold of *empirical presentations* first of all brings into legitimate connection, by which it can become *empirical* KNOWLEDGE, i.e. experience. These mathematical physicists were thus quite unable to dispense with such metaphysical principles, and amongst them, not even with that which makes the conception of their own special subject, namely, matter, available *à priori,* in its application to external experience (as the conception of motion, of the filling of space, of inertia, etc.). But to allow merely empirical principles to obtain in such a question, they rightly held as quite unsuited to the apodictic certainty they desired to give to their natural laws, and hence they preferred to postulate such, without investigating their sources *à priori.*

But it is of the utmost importance in the progress of the sciences, to sever heterogeneous principles from one another, to bring each into a special system, so that it may constitute a science of its own kind, and thereby to avoid the uncertainty springing from their confusion, owing to our not being able to distinguish to which of the two, on the one hand the limitations, and on the other the mistakes occurring in their use, are to be attributed. For this reason I have regarded it as necessary to present in one system the first principles of the pure portion of natural science (*physica generalis*) where mathematical constructions traverse one another, and at the same time the principles of the construction of these conceptions; in short, the possibility of a mathematical doctrine of nature itself. This separation, besides the uses already mentioned, has the special charm, which the unity of knowledge brings with it, if we take care that the boundaries of the sciences do not run into one another, but occupy properly their subdivided fields.

It may serve as a second ground for gauging this procedure, that in all that is called metaphysics the *absolute completeness* of the sciences may be hoped for, in such a manner as can be promised by no other species of knowledge, and therefore, just as in the metaphysics of nature generally, so here also, the completeness of corporeal nature may be confidently expected; the reason being, that in metaphysics the object is considered merely according to the universal laws of thought, but in other sciences as it must be presented according to *data* of intuition (empirical as well as pure). Hence the former, because the object must be invariably compared *with all* the necessary laws of thought, must furnish a *definite* number of cognitions, which can be fully exhausted; but the latter, because it offers an endless multiplicity of intuitions (pure or empirical), and therefore of objects of thought, can never attain to absolute completeness, but can be extended to infinity, as in pure mathematics and empirical natural knowledge. This metaphysical corporeal doctrine I believe myself to have, as far as it reaches, completely exhausted, but do not affect thereby to have achieved any great work.

The scheme for the completeness of a metaphysical system, whether of nature in general, or of corporeal nature in particular, is the table of the categories. For there are not any more pure conceptions of the Understanding, which concern the nature of things. Under the four classes of Quantity, Quality, Relation, and finally Modality, all the determinations of the universal conception of a matter in general, and, therefore, of all that can be thought *à priori* respecting it, that can be presented in mathematical construction, or given in experience as its definite object, must be capable of being brought. There is no more to do in the way of discovery or addition, although certainly, should there be anything lacking in clearness or thoroughness, it may be made better.

Hence the conception of matter had to be carried out through all the four functions of the conceptions of the the understanding (in four divisions), in each of which a new determination of the same was added. The fundamental determination of a something that is to be an object of the external sense, must be motion, for thereby only can this sense be affected. The understanding leads all other predicates pertaining to the nature of matter back to this, and thus natural science is throughout either a pure or an applied *doctrine of motion.* The *metaphysical* foundations of natural science may thus be brought under *four* main divisions, of which the *first* – *motion* considered as pure *quantum,* according to its composition, without any *quality* of the movable, may be termed PHORONOMY; the *second,* which regards it as belonging to the *quality* of the matter, under the name of an original moving force, may be called DYNAMICS; and the *third,* where matter with this quality is conceived as by its own reciprocal motion in *relation,* appears under the name of MECHANICS; and the *fourth,* where its motion or rest [is conceived], merely in reference to the mode of presentation or *modality,* in other words as determined as phenomenon of the external sense, is called PHENOMENOLOGY.

But besides the above internal necessity, whereby the metaphysical foundations of the doctrine of body are not only to be distinguished from physics, which employs empirical principles, but even from the rational premises of the latter, in which the employment of mathematics is to be met with, there is an external, and, though only accidental, at the same time an important reason, for separating its thorough working-out from the general system of metaphysics, and for presenting it systematically as a special whole. For if it be permissible to indicate the boundaries of a science, not merely according to the construction of its object, and its specific kind of cognition, but also according to the aim that is kept in view as a further use of the science itself, and it is found that metaphysics has engaged so many heads, and will continue to engage them, not in order to extend natural knowledge (which could be done much more easily and certainly by observation, experiment, and the application of mathematics to external phenomena), but in order to attain to a knowledge of that which lies wholly beyond all the boundaries of experience, of God, Freedom, and Immortality; [in this case] one gains in the promotion of this object, if one liberates it from a shoot springing indeed from its own stem, but only detrimental to its regular growth, and plants this [shoot] apart, without thereby mistaking its origination, or ignoring its entire growth from the system of general metaphysics. This does not affect the completeness of the latter, but it facilitates the uniform progress of this science towards its goal, if in all cases where the universal doctrine of body is required, one can call to aid the separate system of such a science, without encumbering it with the larger system [viz. of metaphysics in general]. It is indeed very remarkable (though it cannot here be thoroughly entered into), that universal metaphysics, in all cases where it requires instances (intuitions) to procure significance for its pure conceptions of the understanding, must always take them from the universal doctrine of body; in other words, from the form and principle of external intuition; and if these are not found to hand in their entirety, it gropes uncertainly and tremblingly amid mere empty conceptions. Hence the well-known disputes, or at least the obscurity in questions, as to the possibility of an opposition of realities, of intensive quantity, &c., by which the understanding is only taught, through instances from corporeal nature, what the conditions are under which the above conceptions can alone have objective reality, that is, significance and truth. And thus a separate metaphysics of corporeal nature does excellent and indispensable service to the *universal* [metaphysics], in that it procures instances (cases *in concreto*) in which to realise the conceptions and doctrines of the latter (properly the transcendental philosophy), that is, to give to a mere form of thought sense and meaning.

I have in this treatise followed the mathematical method, if not with all strictness (for which more time would have been necessary than I had to devote to it), at least imitatively, not in order, by a display of profundity, to procure a better reception for it, but because I believe such a system to be quite capable of it, and that perfection may in time be obtained by a cleverer hand, if stimulated by this sketch, mathematical investigators of nature should find it not unimportant to treat the metaphysical portion, which anyway cannot be got rid of, as a special fundamental department of general physics, and to bring it into unison with the mathematical doctrine of motion.

Newton, in the preface to his mathematical principles of natural science (after having remarked that geometry only requires two of the mechanical actions which it postulates, namely, to describe a straight line and a circle) says: *geometry is proud of being able to achieve so much while taking so little from extraneous sources.*[2] One might say of metaphysics, on the other hand: *it stands astonished, that with so much offered it by pure mathematics it can effect so little.* In the meantime, this little is something which mathematics indispensably requires in its application to natural science, which, inasmuch as it must here necessarily borrow from metaphysics, need not be ashamed to allow itself to be seen in company with the latter.

Notes

1 Essence is the primal inner principle of all that belongs to the possibility of a thing. Hence one can only predicate an essence, but not a nature of geometrical figures (for nothing is contained in their conception expressive of an existence).

2 Gloria geometria, quod tam paucis principiis alicunde petitis tam multa praestet. – *Newton,* Princ. Phil. Nat. Math. Praefat.

Editorial Headnote

5. Thomas Reid, *Essays on the Intellectual Powers of Man* (1786)

The philosophical approach of Thomas Reid (1710–1796) departs from a careful consideration of the fundamental principles or *axioms* of our sciences. Rather than being the result of philosophical analysis or intuition (as Kant would have it, for example), or deductive necessity (as might be the case in mathematics), for Reid, the statements that we call *a priori* are merely those principles which are self-evident, about which we have never needed to be taught, and to which we would all instantly agree – those for which we would say someone was "destitute of common sense" (p. 43) if they attempted to dispute them. This makes Reid, with other authors like Dugald Stewart (1753–1828), one of the leading figures in what has since been called the "Scottish common sense" school of philosophy.

The philosophical status of Reid's axioms or first principles, on the contrary, is a matter of some dispute. One might say, emphasizing the *evident* in self-evident, that they are principles of evidence: they are principles that can be the object of rational or justified belief without further input from the external world. Or, one might say that these principles guarantee their own *truth*: it is sufficient merely to consider such a first principle to know that it is, in fact, a true statement.[1] Which of these interpretations Reid intended is not obvious from the extant textual evidence, though the question has important philosophical consequences.

Let's return to how Reid's philosophy might be applied to a mature science like Newtonian mechanics. How do we derive principles of sound scientific methodology from common sense? They seem to go far beyond the more straightforward self-evidence of something like simple arithmetic. Here we have yet another fundamental worry about Newtonian methodology. Following Newton's success, we would hope that he had correctly discovered where first principles are to be found in mechanics – that is, that he knew the correct place at which to accept physical principles as fundamental and no longer in need of further argument.

But how could we defend such a choice? What makes a first principle a first principle, or something so evident that we have no need to argue in favor of its truth?

In answering this question, Reid spares no vitriol for those who would create "conjectures," "theories," or "hypotheses." In stark contradiction to Émilie du Châtelet's insistence on the importance of the hypothetical method in the history of science, Reid claims that real scientific discoveries "have always been made by patient observation [or] by accurate experiments . . . and such discoveries have always tended to refute, but not to confirm, the theories and hypotheses which ingenious men had invented" (p. 51). For Reid, this is a straightforward result of the fact that human wisdom and craftsmanship are incalculably inferior compared to divine wisdom and craftsmanship. This leads, naturally enough, to a radically empiricist orientation in Reid's philosophy. He writes:

> Let us accustom ourselves to try every opinion by the touchstone of fact and experience. What can fairly be deduced from facts duly observed, or sufficiently attested, is genuine and pure; it is the voice of God, and no fiction of human imagination. (p. 52)

Interestingly, Reid, too, hopes to enlist Newton on his side in treatment of hypotheses "with just contempt, as the reveries of vain and fanciful men" (p. 51).[2] He quotes, in Latin, from the General Scholium to the *Principia*, where Newton wrote that

> I have not as yet been able to deduce from phenomena the reason for these properties of gravity, and I do not feign hypotheses. For whatever is not deduced from the phenomena must be called a hypothesis; and hypotheses, whether metaphysical or physical, or based on occult qualities, have no place in experimental philosophy.

Reid's presentation of what he takes to be "Newtonian" methodological tenets stands as one of the clearest and most outspoken defenses of the anti-hypothetical reading of Newton to be found in eighteenth-century philosophy.

Notes

1 J. Van Cleve, 'Reid on the First Principles of Contingent Truths,' *Reid Studies* 3:1 (1999), pp. 3–30.

2 For more on other aspects of Reid's Newtonianism, see R. Nichols and G. Yaffe, 'Thomas Reid,' in E. N. Zalta and U. Nodelman (eds), *The Stanford Encyclopedia of Philosophy* (Summer 2023 Edition), at https://plato.stanford.edu/archives/sum2023/entries/reid/ [accessed 16 July 2024].

5

ESSAYS ON THE INTELLECTUAL POWERS OF MAN

Thomas Reid

Source: Thomas Reid, *Essays on the Intellectual Powers of Man* (1786), pp. 1:33–1:52.

CHAP. II.

Principles taken for granted.

AS there are words common to Philosophers and to the vulgar, which need no explication; so there are principles common to both, which need no proof, and which do not admit of direct proof.

One who applies to any branch of science must be come to years of understanding, and consequently must have exercised his reason, and the other powers of his mind, in various ways. He must have formed various opinions and principles by which he conducts himself in the affairs of life. Of those principles, some are common to all men, being evident in themselves, and so necessary in the conduct of life, that a man cannot live and act according to the rules of common prudence without them.

All men that have common understanding agree in such principles, and consider a man as lunatic or destitute of common sense, who denies, or calls them in question. Thus, if any man were found of so strange a turn as not to believe his own eyes; to put no trust in his senses, nor have the least regard to their testimony; would any man think it worth while to reason gravely with such a person, and, by argument, to convince him of his error! Surely no wise man would. For before men can reason together, they must agree in first principles; and it is impossible to reason with a man who has no principles in common with you.

There are, therefore, common principles, which are the foundation of all reasoning, and of all science. Such common principles seldom admit of direct proof, nor do they need it. Men need not to be taught them; for they are such as all men of common understanding know; or such, at least, as they give a ready assent to, as soon as they are proposed and understood.

Such principles, when we have occasion to use them in science, are called *axioms*. And, although it be not absolutely necessary, yet it may be of great use, to point out the principles or axioms on which a science is grounded.

DOI: 10.4324/9781003263814-6

Thus, mathematicians, before they prove any of the propositions of mathematics, lay down certain axioms, or common principles, upon which they build their reasonings. And although those axioms be truths which every man knew before; such as, That the whole is greater than a part, That equal quantities added to equal quantities, make equal sums; yet, when we see nothing assumed in the proof of mathematical propositions, but such self-evident axioms, the propositions appear more certain, and leave no room for doubt to dispute.

In all other sciences, as well as in mathematics, it will be found, that there are a few common principles, upon which all the reasonings in that science are grounded, and into which they may be resolved. If these were pointed out and considered, we should be better able to judge what stress may be laid upon the conclusions in that science. If the principles be certain, the conclusions justly drawn from them must be certain. If the principles be only probable, the conclusions can only be probable. If the principles be false, dubious, or obscure, the superstructure that is built upon them must partake of the weakness of the foundation.

Sir ISAAC NEWTON, the greatest of Natural Philosophers, has given an example well worthy of imitation, by laying down the common principles or axioms, on which the reasonings in natural philosophy are built. Before this was done, the reasonings of Philosophers, in that science, were as vague and uncertain as they are in most others. Nothing was fixed; all was dispute and controversy: But, by this happy expedient, a solid foundation is laid in that science, and a noble superstructure is raised upon it, about which there is now no more dispute or controversy among men of knowledge, than there is about the conclusions of mathematics.

It may, however, be observed, that the first principles of natural philosophy are of a quite different nature from mathematical axioms: They have not the same kind of evidence, nor are they necessary truths, as mathematical axioms are: They are such as these: That similar effects proceed from the same or similar causes: That we ought to admit of no other causes of natural effects, but such as are true, and sufficient to account for the effects. These are principles, which, though they have not the same kind of evidence that mathematical axioms have; yet have such evidence, that every man of common understanding readily assents to them, and finds it absolutely necessary to conduct his actions and opinions by them, in the ordinary affairs of life.

Though it has not been usual, yet, I conceive it may be useful, to point out some of those things which I shall take for granted, as first principles in treating of the mind and its faculties. There is the more occasion for this; because very ingenious men, such as DES CARTES, MALEBRANCHE, ARNAULD, LOCKE, and many others, have lost much labour, by not distinguishing things which require proof, from things which, though they may admit of illustration, yet being self-evident, do not admit of proof. When men attempt to deduce such self-evident principles from others more evident, they always fall into inconclusive reasoning: And the consequence of this has been, that others, such as BERKELEY and HUME, finding the arguments brought to prove such first principles to be weak and inconclusive, have been tempted first to doubt of them, and afterwards to deny them.

It is so irksome to reason with those who deny first principles, that wise men commonly decline it. Yet it is not impossible, that what is only a vulgar prejudice may be mistaken for a first principle. Nor is it impossible, that what is really a first principle may, by the enchantment of words, have such a mist thrown about it, as to hide its evidence, and to make a man of candour doubt of it. Such cases happen more frequently perhaps in this science than in any other; but they are not altogether without remedy. There are ways by which the evidence of first principles may be made more apparent when they are brought into dispute; but they require to be handled in a way peculiar to themselves. Their evidence is not demonstrative, but intuitive. They require not proof, but to be placed in a proper point of view. This will be shown more fully in its proper place, and applied to those very principles which we now assume. In the mean time, when they are proposed as first principles, the reader is put on his guard, and warned to consider whether they have a just claim to that character.

1. *First,* then, I shall take it for granted, that I *think,* that I *remember,* that I *reason,* and, in general, that I really perform all those operations of mind of which I am conscious.

The operations of our minds are attended with consciousness; and this consciousness is the evidence, the only evidence which we have or can have of their existence. If a man should take it into his head to think or to say that his consciousness may deceive him, and to require proof that it cannot, I know of no proof that [can] be given him; he must be left to himself as a man that denies first principles, without which there can be no reasoning. Every man finds himself under a necessity of believing what consciousness testifies, and every thing that hath this testimony is to be taken as a first principle.

2. As by consciousness we know certainly the existence of our present thoughts and passions; so we know the past by remembrance. And when they are recent, and the remembrance of them fresh, the knowledge of them, from such distinct remembrance, is, in its certainty and evidence, next to that of consciousness.

3. But it is to be observed, that we are conscious of many things to which we give little or no attention. We can hardly attend to several things at the same time; and our attention is commonly employed about that which is the object of our thought, and rarely about the thought itself. Thus, when a man is angry, his attention is turned to the injury done him, or the injurious person; and he gives very little attention to the passion of anger, although he is conscious of it. It is in our power, however, when we come to the years of understanding, to give attention to our own thoughts and passions, and the various operations of our minds. And when we make these the objects of our attention, either while they are present, or when they are recent and fresh in our memory, this act of the mind is called *reflection.*

We take it for granted, therefore, that, by attentive reflection, a man may have a clear and certain knowledge of the operations of his own mind; a knowledge no less clear and certain, than that which he has of an external object when it is set before his eyes.

This reflection is a kind of intuition, it gives a like conviction with regard to internal objects, or things in the mind, as the faculty of seeing gives with regard to objects of sight. A man must, therefore, be convinced beyond possibility of doubt, of every thing with regard to the operations of his own mind, which he clearly and distinctly discerns by attentive reflection.

4. I take it for granted that all the thoughts I am conscious of, or remember, are the thoughts of one and the same thinking principle, which I call *myself,* or my *mind.* Every man has an immediate and irresistible conviction, not only of his present existence, but of his continued existence and identity, as far back as he can remember. If any man should think fit to demand a proof that the thoughts he is successively conscious of belong to one and the same thinking principle. If he should demand a proof that he is the same person today as he was yesterday, or a year ago, I know no proof that can be given him: He must be left to himself, either as a man that is lunatic, or as one who denies first principles, and is not to be reasoned with.

Every man of a sound mind finds himself under a necessity of believing his own identity, and continued existence. The conviction of this is immediate and irresistible; and if he should lose this conviction, it would be a certain proof of insanity, which is not to be remedied by reasoning.

5. I take it for granted that there are some things which cannot exist by themselves, but must be in something else to which they belong, as qualities or attributes.

Thus motion cannot exist but in something that is moved. And to suppose that there can be motion while every thing is at rest, is a gross and palpable absurdity. In like manner, hardness and softness, sweetness and bitterness, are things which cannot exist by themselves; they are qualities of something which is hard or soft, sweet or bitter: That thing, whatever it be, of which they are qualities, is called their subject, and such qualities necessarily suppose a subject.

Things which may exist by themselves, and do not necessarily suppose the existence of any thing else, are called substances; and with relation to the qualities or attributes that belong to them, they are called the subjects of such qualities or attributes.

All the things which we immediately perceive by our senses, and all the things we are conscious of, are things which must be in something else as their subject. Thus by my senses, I perceive figure, colour, hardness, softness, motion, resistance and such like things. But these are qualities, and must necessarily be in something that is figured, coloured, hard or soft, that moves, or resists. It is not to these qualities, but to that which is the subject of them, that we give the name of body. If any man should think fit to deny that these things are qualities, or that they require any subject, I leave him to enjoy his opinion as a man who denies first principles, and is not fit to be reasoned with. If he has common understanding, he will find that he cannot converse half an hour without saying things which imply the contrary of what he professes to believe.

In like manner, the things I am conscious of, such as thought, reasoning, desire, necessarily suppose something that thinks, that reasons, that desires. We do not give the name of mind to thought, reason, or desire; but to that being which thinks, which reasons, and which desires.

That every act or operation, therefore, supposes an agent, that every quality supposes a subject, are things which I do not attempt to prove, but take for granted. Every man of common understanding discerns this immediately, and cannot entertain the least doubt of it. In all languages we find certain words which, by Grammarians, are called adjectives. Such words denote attributes, and every adjective must have a substantive to which it belongs; that is, every attribute must have a subject. In all languages we find active verbs which denote some action or operation; and it is a fundamental rule in the grammar of all languages, that such a verb supposes a person; that is, in other words, that every action must have an agent. We take it, therefore, as a first principle, that goodness, wisdom, and virtue, can only be in some being that is good, wise, and virtuous; that thinking supposes a being that thinks; and that every operation we are conscious of supposes an agent that operates, which we call mind.

6. I take it for granted that, in most operations of the mind, there must be an object distinct from the operation itself. I cannot see, without seeing something. To see without having any object of sight is absurd. I cannot remember, without remembering something. The thing remembered is past, while the remembrance of it is present; and therefore the operation and the object of it must be distinct things. The operations of our minds are denoted, in all languages, by active transitive verbs, which, from their construction in grammar, require not only a person or agent, but likewise an object of the operation. Thus the verb know, denotes an operation of mind. From the general structure of language, this verb requires a person; I know, you know, or he knows: But it requires no less a noun in the accusative case, denoting the thing known; for he that knows, must know something; and to know, without having any object of knowledge, is an absurdity too gross to admit of reasoning.

7. We ought likewise to take for granted, as first principles, things wherein we find an universal agreement, among the learned and unlearned, in the different nations and ages of the world. A consent of ages and nations, of the learned and vulgar, ought, at least, to have great authority, unless we can show some prejudice, as universal as that consent is, which might be the cause of it. Truth is one, but error is infinite. There are many truths so obvious to the human faculties, that it may be expected that men should universally agree in them. And this is actually found to be the case with regard to many truths, against which we find no dissent, unless perhaps that of a few sceptical Philosophers, who may justly be suspected, in such cases, to differ from the rest of mankind, through pride, obstinacy, or some favourite passion. Where there is such universal consent in things not deep nor intricate, but which lie, as it were, on the surface, there is the greatest presumption that can be, that it is the natural result of the human faculties; and it must have great authority with every sober mind that loves truth. *Major enim pars eo fere deferri solet quo a natura deducitur.* Cic. de Off. 1. 41.

Perhaps it may be thought, that it is impossible to collect the opinions of all men upon any point whatsoever, and, therefore, that this maxim can be of no use. But there are many cases wherein it is otherwise. Who can doubt, for instance, whether mankind have, in all ages, believed the existence of a material world, and that those things which they see and handle are real, and not mere illusions and apparitions? Who can doubt, whether mankind have universally believed, that every thing that begins to exist, and every change that happens in nature, must have a cause? Who can doubt, whether mankind have been universally persuaded that there is a right and a wrong in human conduct? Some things which, in certain circumstances, they ought to do, and other things which they ought not to do? The universality of these opinions, and of many such that might be named, is sufficiently evident, from the whole tenor of mens conduct, as far as our acquaintance reaches, and from the records of history, in all ages and nations, that are transmitted to us.

There are other opinions that appear to be universal, from what is common in the structure of all languages, ancient and modern, polished and barbarous. Language is the express image and picture of human thoughts; and, from the picture, we may often draw very certain conclusions with regard to the original. We find in all languages the same parts of speech, nouns substantive and adjective, verbs active and passive, varied according to the tenses of past, present, and future; we find adverbs, prepositions, and conjunctions. There are general rules of syntax common to all languages. This uniformity in the structure of language, shows a certain degree of uniformity in those notions upon which the structure of language is grounded.

We find, in the structure of all languages, the distinction of acting and being acted upon, the distinction of action and agent, of quality and subject, and many others of the like kind; which shows, that these distinctions are founded in the universal sense of mankind. We shall have frequent occasion to argue from the sense of mankind expressed in the structure of language; and therefore it was proper here to take notice of the force of arguments drawn from this topic.

8. I need hardly say, that I shall also take for granted such facts as are attested to the conviction of all sober and reasonable men, either by our senses, by memory, or by human testimony. Although some writers on this subject have disputed the authority of the senses, of memory, and of every human faculty; yet we find, that such persons, in the conduct of life, in pursuing their ends, or in avoiding dangers, pay the same regard to the authority of their senses, and other faculties, as the rest of mankind. By this they give us just ground to doubt of their candour in their professions of sceptiscism.

This, indeed, has always been the fate of the few that have professed scepticism, that, when they have done what they can to discredit their senses, they find themselves, after all, under a necessity of trusting to them. Mr. Hume has been so candid as to acknowledge this; and it is no less true of those who have not shown the same candour: For I never heard that any sceptic run his head against a post, or stept into a kennel, because he did not believe his eyes.

Upon the whole, I acknowledge that we ought to be cautious, that we do not adopt opinions as first principles, which are not entitled to that character. But there is surely the least danger of men being imposed upon in this way, when such principles openly lay claim to the character, and are thereby fairly exposed to the examination of those who may dispute their authority. We do not pretend, that those things that are laid down as first principles may not be examined, and that we ought not to have our ears open to what may be pleaded against their being admitted as such. Let us deal with them, as an upright judge does with a witness who has a fair character. He pays a regard to the testimony of such a witness, while his character is unimpeached. But if it can be shown that he is suborned, or that he is influenced by malice or partial favour, his testimony loses all its credit, and is justly rejected.

CHAP. III.

Of Hypotheses.

EVERY branch of human knowledge hath its proper principles, its proper foundation and method of reasoning; and, if we endeavour to build it upon any other foundation, it will never stand firm and stable. Thus the historian builds upon testimony, and rarely indulges conjecture. The antiquarian mixes conjecture with testimony; and the former often makes the larger ingredient. The mathematician pays not the least regard either to testimony or conjecture, but deduces every thing, by demonstrative reasoning, from his definitions and axioms. Indeed, whatever is built upon conjecture, is improperly called science; for conjecture may beget opinion, but cannot produce knowledge. Natural philosophy must be built upon the phænomena of the material system, discovered by observation and experiment.

When men first began to philosophise, that is, to carry their thoughts beyond the objects of sense, and to enquire into the causes of things, and the secret operations of nature, it was very natural for them to indulge conjecture; nor was it to be expected, that, in many ages, they should discover the proper and scientific way of proceeding in philosophical disquisitions. Accordingly we find, that the most ancient systems in every branch of philosophy were nothing but the conjectures of men famous for their wisdom, whose fame gave authority to their opinions. Thus, in early ages, wise men conjectured, that this earth is a vast plain, surrounded on all hands by a boundless ocean. That from this ocean, the sun, moon, and stars, emerge at their rising, and plunge into it again at their setting.

With regard to the mind, men in their rudest state are apt to conjecture, that the principle of life in a man is his breath; because the most obvious distinction between a living and a dead man is, that the one breathes, and the other does not. To this it is owing, that, in ancient languages, the word which denotes the soul, is that which properly signifies breath or air.

As men advance in knowledge, their first conjectures appear silly and childish, and give place to others, which tally better with later observations and discoveries. Thus one system of philosophy succeeds another, without any claim to superior merit; but this, that it is a more ingenious system of conjectures, and accounts better for common appearances.

To omit many ancient systems of this kind, DES CARTES, about the middle of the last century, dissatisfied with the *materia prima,* the *substantial forms,* and the *occult qualities* of the Peripatetics, conjectured boldly, that the heavenly bodies of our system are carried round by a vortex or whirlpool of subtile matter, just as straws and chaff are carried round in a tub of water. He conjectured, that the soul is seated in a small gland in the brain, called the pineal gland: That there, as in her chamber of presence, she receives intelligence of every thing that affects the senses, by means of a subtile fluid contained in the nerves, called the animal spirits; and that she dispatches these animal spirits, as her messengers, to put in motion the several muscles of the body, as there is occasion. By such conjectures as these, DES CARTES could account for every phænomenon in nature, in such a plausible manner, as gave satisfaction to a great part of the learned world for more than half a century.

Such conjectures in philosophical matters have commonly got the name of *hypotheses,* or *theories.* And the invention of a hypothesis, founded on some slight probabilities, which accounts for many appearances of nature, has been considered as the highest attainment of a Philosopher. If the hypothesis hangs well together, is embellished by a lively imagination, and serves to account for common appearances; it is considered by many as having all the qualities that should recommend it to our belief; and all that ought to be required in a philosophical system.

There is such proneness in men of genius to invent hypotheses, and in others to acquiesce in them, as the utmost which the human faculties can attain in philosophy, that it is of the last consequence to the progress of real knowledge, that men should have a clear and distinct understanding of the nature of hypotheses in philosophy, and of the regard that is due to them.

Although some conjectures may have a considerable degree of probability, yet it is evidently in the nature of conjecture to be uncertain. In every case the assent ought to be proportioned to the evidence; for to believe firmly, what has but a small degree of probability, is a manifest abuse of our understanding. Now, though we may, in many cases, form very probable conjectures concerning the works of men, every conjecture we can form with regard to the works of God, has as little probability as the conjectures of a child with regard to the works of a man.

The wisdom of God exceeds that of the wisest man, more than his wisdom exceeds that of a child. If a child were to conjecture how an army is to be formed in the day of battle; how a city is to be fortified, or a state governed; what chance has he to guess right? As little chance has the wisest man when he pretends to conjecture how the planets move in their courses, how the sea ebbs and flows, and how our minds act upon our bodies.

If a thousand of the greatest wits that ever the world produced, were, without any previous knowledge in anatomy, to sit down and contrive how, and by what internal organs, the various functions of the human body are carried on; how the blood is made to circulate, and the limbs to move, they would not in a thousand years hit upon any thing like the truth.

Of all the discoveries that have been made concerning the inward structure of the human body, never one was made by conjecture. Accurate observations of Anatomists have brought to light innumerable artifices of nature in the contrivance of this machine of the human body, which we cannot but admire as excellently adapted to their several purposes. But the most sagacious physiologist never dreamed of them till they were discovered. On the other hand, innumerable conjectures, formed in different ages, with regard to the structure of the body, have been confuted by observation, and none ever confirmed.

What we have said of the internal structure of the human body, may be said, with justice, of every other part of the works of God, wherein any real discovery has been made. Such discoveries have always been made by patient observation, by accurate experiments, or by conclusions drawn by strict reasoning from observations and experiments; and such discoveries have always tended to refute, but not to confirm, the theories and hypotheses which ingenious men had invented.

As this is a fact confirmed by the history of philosophy in all past ages, it ought to have taught men, long ago, to treat with just contempt hypotheses in every branch of philosophy, and to despair of ever advancing real knowledge in that way. The Indian Philosopher, being at a loss to know how the earth was supported, invented the hypothesis of a huge elephant; and this elephant he supposed to stand upon the back of a huge tortoise. This hypothesis, however ridiculous it appears to us, might seem very reasonable to other Indians, who knew no more than the inventor of it; and the same will be the fate of all hypotheses invented by men to account for the works of God: They may have a decent and plausible appearance to those who are not more knowing than the inventor; but, when men come to be more enlightened, they will always appear ridiculous and childish.

This has been the case with regard to hypotheses that have been revered by the most enlightened part of mankind for hundreds of years; and it will always be the case to the end of the world. For, until the wisdom of men bear some proportion to the wisdom of God, their attempts to find out the structure of his works, by the force of their wit and genius, will be vain.

The finest productions of human art are immensely short of the meanest works of nature. The nicest artist cannot make a feather, or the leaf of a tree. Human workmanship will never bear a comparison with divine. Conjectures and hypotheses are the invention and the workmanship of men, and must bear proportion to the capacity and skill of the inventor; and therefore will always be very unlike to the works of God, which it is the business of philosophy to discover.

The world has been so long befooled by hypotheses in all parts of philosophy, that it is of the utmost consequence to every man, who would make any progress in real knowledge, to treat them with just contempt, as the reveries of vain and fanciful men, whose pride makes them conceive themselves able to unfold the mysteries of nature by the force of their genius. A learned man in an epistle to DES CARTES has the following observation, which very much deserved the attention of that Philosopher, and of all that come after him. "When men, sitting in their closet, and consulting only their books, attempt disquisitions into nature, they may indeed tell how they would have made the world, if God had given them that in commission; that is, they may describe chimeras, which correspond with the imbecillity of their own minds, no less than the admirable beauty of the Universe corresponds with the infinite perfection of its Creator; but without an understanding truly divine, they can never form such an idea to themselves as the Deity had in creating things."

Let us, therefore, lay down this as a fundamental principle in our enquiries into the structure of the mind, and its operations, that no regard is due to the conjectures or hypotheses of Philosophers, however ancient, however generally received. Let us accustom ourselves to try every opinion by the touchstone of fact and experience. What can fairly be deduced from facts duly observed, or sufficiently attested, is genuine and pure; it is the voice of God, and no fiction of human imagination.

The first rule of philosophising laid down by the great NEWTON, is this; *Causas rerum naturalium, non plures admitti debere, quam quæ et veræ sint, et earum phænomenis explicandis sufficiant.* "No more causes, nor any other causes of natural effects ought to be admitted, but such as are both true, and are sufficient for explaining their appearances." This is a golden rule; it is the true and proper test, by which what is sound and solid in philosophy may be distinguished from what is hollow and vain.

If a Philosopher, therefore, pretends to shew us the cause of any natural effect, whether relating to matter or to mind; let us first consider whether there is sufficient evidence that the cause he assigns does really exist. If there is not, reject it with disdain as a fiction which ought to have no place in genuine philosophy. If the cause assigned really exists, consider, in the next place, whether the effect it is brought to explain necessarily follows from it. Unless it has these two conditions, it is good for nothing.

When NEWTON had shown the admirable effects of gravitation in our planetary system, he must have felt a strong desire to know its cause. He could have invented a hypothesis for this purpose, as many had done before him. But his philosophy was of another complexion. Let us hear what he says: *Rationem [vero] harum gravitatis proprietatum ex phænomenis non[dum] potui deducere, et hypotheses non fingo. Quicquid enim ex phænomenis non deducitur hypothesis vocanda est. Et hypotheses, seu metaphysicæ, seu physicæ, seu qualitatum occultarum, seu mechanicæ, in philosophia experimentali locum non habent.*

Editorial Headnote

6. Mary Shepherd, *Essay Upon the Relation of Cause and Effect* (1824)

Lady Mary Shepherd (1777–1847) was among the first women philosophers to work in Scotland. Daughter of a well connected, noble family, her parents' social circle included figures such as William Whewell, Charles Lyell, and Mary Somerville, each of whom we will encounter later in this collection.[1] The book excerpted here, which Shepherd published anonymously in the early nineteenth century, is a long commentary on David Hume's (1711–1776) discussion of causation. (Her later works would be published under her own name.) As with Thomas Reid, whom we saw in the last text and who published several decades later, Hume was also an arch-empiricist, claiming that only that knowledge which could be directly founded upon observed experience was an appropriate ground for belief. Responses, commentaries, and critiques of this tradition formed some of the most fertile ground for epistemological and methodological works in Britain at the turn of the nineteenth century.

As Shepherd and other writers had noticed, this Humean empiricism does not sit well with our everyday causal and scientific commitments. As she quotes, and as is now well known, Hume rejects causation as a really existing relationship in the natural world – since it is impossible for us to *see* a relation of cause and effect, there can be no true knowledge of a connection between causes and their effects. All we have is our habit to draw such a causal link, built after having seen many instances of some event (the cause) followed by many instances of some other event (the effect). Such an interpretation threatens, for instance, the idea that Newton has given us knowledge of the laws that govern the orbit of the planets around the sun.

Shepherd offers us here an *a priori* defense of cause and effect. In doing so, she constructs a definition that resembles what contemporary philosophers of science have called "difference-making" conceptions of causation.[2] If we have a situation in which only one relevant difference gives rise to a change in our perceptions, then we do not need repeated trials, experiments, or "custom" in order to come to believe that the difference is the cause of the perceptual change – "for there is nothing else to make a difference, and a different quality could not '*begin its own existence*'" (p. 56). (This use of a single experiment to infer a causal relationship would later by systematized by both John Herschel and J. S. Mill.) We therefore can argue, *a priori*, that every change in an object must be caused by something else.

But her metaphysical argument moves a step beyond this. If we conceive of objects as being collections of qualities that affect our senses, then changes in these qualities actually amount to the creation of new objects. We can come to understand cause and effect precisely because we can understand the relationships between the *natures* of objects before and after a causal change. Combustion, for example, *just is* one of the properties of an object that is a combination of fire and wood, rather than being the *effect* of such a combination. Such reasoning can also ground our future faith in the continued operation of causal patterns. Causes that are currently alike will produce effects that are currently alike, because if the causes are truly alike, then there is no change in their nature that could give rise to any change in their later effects. If a combination of fire and wood fails to combust, then it must be the case that it in fact isn't a combination of fire and wood at all. We can thus also argue, *a priori*, that the same causes will necessarily continue to produce the same effects.

The importance of such an argument to the defense of scientific methodology comes near the end of the section, when Shepherd argues that, given this picture of the nature of cause and effect, it does not matter whether or not we understand the "secret powers" of bodies. What exactly is it, we might ask, that causes bodies to gravitate toward one another on Newton's theory? On Shepherd's notion of causation, we can offer an argument that this is a meaningless question. What matters is that the qualities of bodies change in the ways described by Newton's laws, and that the differences picked out as important by Newtonian mechanics are in fact responsible for giving rise to the behaviors of physical objects. Which fundamental facts about the nature of matter might give rise to those qualities are not important, she argues, but without the empiricist foundation that had been deployed by Hume or Reid.

Notes

1 D. Boyle, *Mary Shepherd: A Guide* (Oxford: Oxford University Press, 2023), p. 2.
2 J. Wilson, "On Mary Shepherd's *Essay Upon the Relation of Cause and Effect*," in E. Schliesser (ed.), *Neglected Classics of Philosophy, Volume 2* (Oxford: Oxford University Press, 2022), pp. 141–171.

6

ESSAY UPON THE RELATION OF CAUSE AND EFFECT

Mary Shepherd

Source: Mary Shepherd, *Essay Upon the Relation of Cause and Effect* (1824), pp. 40–63.

SECTION THE SECOND.

We will now proceed to the second part of the original inquiry; that is, Why "we conclude that such particular Causes must necessarily have such particular Effects; and what is the nature of that inference we draw from one to the other, and of the belief we repose in it? The question, however, ought to stand thus, "why LIKE CAUSES must necessarily have LIKE EFFECTS? because what is really enquired into, is the *general notion of necessary connexion,* between *all like* Cause and Effect; and by thus putting the question respecting *particulars only,* although they might be included in an universal answer, yet no answer applicable to them MERELY, could authorize an *universal axiom.* The manner of stating the enquiry in *the Essays,* is also too vaguely expressed, (although it be evident that it is the *general relation* which is enquired into.) Mr. Hume says, "we will now enquire, how we arrive at the *knowledge of Cause and Effect.*"[1] *It ought to be stated,* how we arrive at the knowledge of the *necessary connexion,* between *like* Cause and Effect?

Let it be remembered, that Mr. Hume says, "this principle is nothing but custom and habit;" that "belief in necessary connexion is nothing but an intense and steady conception, arising from the customary conjunction of the object with something present to the memory or senses; that when flame and heat, cold and snow, have always been conjoined together, there is such a customary conjunction between them, that when flame and snow are anew presented to the senses, the mind is carried by custom to expect heat and cold."

"That *reason* can never show us the connexion of one object with another, though aided by experience; for we can at least *conceive a change in the course of nature.* That necessary connexion is nothing but an internal act of the mind, determined to carry its thoughts from one object to another." Thus *necessary connexion* of cause and effect is only a custom of the mind! *Power* is only a custom of the mind! Expectations, and experience, are only customs of the mind! The consequence of which doctrine is, that as a *custom of the mind* is entirely a different circumstance from the *operation of nature,* we may "*conceive*" at least the

DOI: 10.4324/9781003263814-7

contrary of what we have been accustomed to may take place, – we may conceive the "course of nature to change."

Now it is my intention to shew, in contradiction to these ideas of Mr. Hume, that it is *Reason,* and not *Custom,* which guides our minds in forming the notions of necessary connexion, of belief and of expectation.[2]

In order to this let us bear in mind the reasoning already adduced in the foregoing Chapter, and it thence immediately follows, that objects which we know by our senses do begin their existences, and by our reason know they cannot begin it of themselves, must begin it by the operation of some *other beings* in existence, producing these new qualities in nature, and introducing them to our observation. The very meaning of the word Cause, is *Producer* or *Creator;* of Effect, the *Produced* or *Created* – and the idea is gained by such an observance of nature, as we think is efficient in any given case, to an *experimentum crucis.*

Long observation of the invariableness of antecedency, and subsequency, is not wanted; many trials are not wanted, to generate the notion of *producing power.*

One trial is enough, in such circumstances, as will bring the mind to the following reasoning.

Here is a new quality, which appears to my senses:

But it could not arise of itself; nor could any surrounding objects, but one (or more) affect it; therefore that one, (or more) have occasioned it, for there is nothing else to make a difference; and a *difference* could not "*begin of itself.*"

This is an argument, which all persons, however illiterate, feel the force of. It is the only foundation for the demonstrations of the laboratory of the chymist; which all life resembles, and so closely, in many instances, that the philosopher, and the vulgar, are equally sure of what cause is absolutely necessary to the production of certain effects; for instance, each knows that in certain given circumstances, *the closing of the Eye* will eclipse the prospect of nature; and the slight motion of reopening it, will restore all the objects to view. Therefore, the Eye (in these circumstances,) is the *Cause* or *Producer of vision.* ONE trial would be enough, under certain *known* circumstances.[3] Why? not from "*custom,*" because there has been *one trial only;* but from *Reason,* because vision not being able *to produce itself, nor any of the surrounding objects by the supposition;* it is the *Eye* which must necessarily perform the operation; for there is nothing else to make a difference; and a different quality could not "*begin its own existence.*" It is this sort of REASONING UPON EXPERIMENT, which takes place in every man's mind, concerning every affair in life, which generates the notion of Power, and necessary Connexion; and gives birth to that maxim, "*a like Cause must produce a like Effect.*" The circumstances being supposed the same on a second occasion as on a former one, and carefully observed to be so; the Eye when opened would be expected to let in light, and all her objects. "I observe (says the mind) in this or any other case, all the prevening circumstances the same as before; for there is nothing to make a difference; and a difference cannot arise without something to occasion it; else there would be a *beginning of existence* by itself, which is impossible."

It is this compound idea, therefore, *the result of the experience of what does take place upon any given trial,* MIXED with the *reasoning that nothing else could ensue,* unless on the one hand, *efficient causes were allowed for the alteration;* or, on the other, that things could "*alter their existences* FOR THEMSELVES;" which generates the notion of *power or "producing principle," and for which we have formed the word.*

It is in vain to say that a habit of association of ideas from observing "*contiguity in time, and place,*" between objects is all we know of *power;* a habit of the mind will not *begin existence,* will not *introduce a quality.* The really philosophical method of viewing the subject is this: that objects in relation to us, are nothing but masses of certain qualities, affecting certain of our senses; and which, when independent of our senses, are *unknown* powers or qualities in nature. These masses change their qualities by their mixture with any other mass, and then the corresponding qualities determined to the senses must of course also change. These changed qualities, are termed *effects;* or *consequents;* but are really no more than NEW QUALITIES arising from *new objects,* which have been formed by the *junctions of other objects* (previously formed) or might be considered as the *unobserved qualities* of *existing objects;* which *shall be observed when properly exhibited.*

If then an existence now in being, *conjoined with any other,* forms thereby *a new nature,* capable of exhibiting *new qualities,* these new qualities must enter into the definition of the objects; they become a part of their natures; and when by careful experiment, or judicious observation, no new prevening circumstances are supposed to make an alteration in the conjunction of the same bodies, the *new qualities,* that are named *effects,* are expected without a doubt to arise upon every such conjunction; because, they as much belong to this *newly combined nature,* as the original qualities did to each separate nature, before their conjunction. So little is custom the principle of cause and effect, that if upon the *first* and original trial of the element of fire, all surrounding circumstances were put away from having any influence over it, saving the body it destroyed; that power of *discerptibility* would be ever after considered as one of its qualities; as much as its colour or its light, or its warmth, without the presence of which, it would not be fire.

This conjunction with a grosser material than itself, is the new circumstance, on which it exhibits its essential and permanent quality of discerptibility to the senses; now if the trial be complete, when upon a second occasion an object having the same sensible qualities as fire hath, known also to have been elicited from the same prevening circumstances, meets with the same gross body as heretofore, it must of *necessity* consume it. There is nothing to make a *difference.* A *difference* is an *Effect,* a *change of being,* an *altered existence,* an existence which *cannot "begin* of itself" any more than any other in Nature; could the fire be supposed not to consume the gross body, there would be a *difference* of qualities, that is, new qualities, which by the data there is no cause for. The original circumstances, of which fire is the compound Effect, from which it results as a *formed object,* are supposed to be ordered the same as on a former occasion; these are necessarily compelled to be attended with the same effects or combined qualities; otherwise there would be the "*beginnings of existence*" by themselves, which has before been shown to be impossible. But the *combined qualities,* are the whole qualities that fire in every circumstance, is capable of producing. Meeting, therefore, with a gross body, which on any one occasion, in certain circumstances, it once consumed; under the same circumstances, it must necessarily again consume it. That DIFFERENCES OF EXISTENCE cannot begin of themselves; is therefore the second conclusion supposed to be established.

"*Antecedency* and *subsequency,*" are therefore immaterial to the proper definition of Cause and Effect;" on the contrary, although an object, in order to act as a Cause, must be in Being antecedently to such action; yet when it *acts as a Cause,* its *Effects* are *synchronous with that action,* and are *included in it;* which a close inspection into the nature of cause will prove. For effects are no more than the new qualities, of newly formed objects. Each conjunction of bodies, (now separately in existence, and of certain defined qualities,) produces upon their union those new natures, whose qualities must necessarily *be in,* and *with them, in the very moment of their formation.*

Thus *the union of two distinct natures,* is the *cause, producer* or *creator* of another; which must instantly, and immediately, have all its peculiar qualities; but the cause has not acted, is not completed, till the *union* has taken place, and the new nature is formed with all its qualities, *in,* and *about it. Cause producing Effect,* therefore, under the strict eye of philosophical scrutiny, is a *new object* exhibiting *new qualities;* or shortly, the formation of a new *mass of qualities.* A *chain of conjunctions of bodies,* of course, *occupies time;* and is the reason why the careless observation of philosophers, enabling them to take notice only of some one distinct effect, (after perhaps innumerable successive conjunctions of bodies,) occasions the mistake, by which they consider *subsequency of effect,* as a part of the *essential definition* of that term; and *priority,* as *essential* to the nature of Cause.

As a short illustration of the doctrine unfolded, let us take the idea of nourishment, considered as the effect, subsequent to the taking of food, its cause. Here the *nature* of nourishment, is *a process* which begins to act immediately that food is in conjunction with the stomach. "That we are nourished;" is only the last result of a continuous chain of causes and effects, in formation from the first moment the food enters the stomach, to that, in which every particle is absorbed and deposited in the proper place in the body. Here, the capacity of food to exhibit certain qualities, when in conjunction with the body, is shown; the nature of the human body, to exhibit certain other qualities, in consequence of that conjunction, is also shown; but the *effect of nourishment,* being *subsequent* to, and at such a distance of time from, the original Cause, is only so, on account of its being the effect of a vast number of causes, or unions of objects in succession, of which the union of the stomach and the food was first in order.

Our deficient observation, is apt to prevent our taking notice of the 2d, 3d, or indefinite number of effects; which arise in consequence of as many conjunctions of objects.

But the first, and other *effects* successively, are as much and entirely synchronous with their *causes,* as any other quality of any single object, which is always exhibited along with it.

2dly. It is also quite immaterial to the definition of this relation, whether an untried, or unobserved quality, be called *quality,* or *effect*. The unknown or at present undetermined quality, which is termed an effect, might always change its place with some *known quality,* and not bear the name of effect; and *vice versa:* Thus, a blind man may call the object which warmed, or burned him, fire; but his eyes being supposed suddenly to open, he would consider the flame and its brilliant colour as the *effects* of fire; whilst he who sees fire constantly, being able always to take notice of its flame and colour, considers them as the constant and unvarying *qualities* of fire, and which render the substance before him worthy of bearing that name; but the quality of burning, which he does not *constantly* experience, he names an *effect* or *consequence* of fire previously being in existence. But the true method of looking upon the subject is this – that fire, in order to deserve the name it bears, must comprehend all its qualities *tried* and *untried; observed* and *unobserved; determined* and *undetermined;* it deserves the name only on account of its being a certain defined object; *elicited from certain causes observed to be efficient to its production;* and by the very conditions of the question, is allowed to be *the same.* But an object is nothing else (in relation to *us,*) than a mass of peculiar qualities; and when observations inform us, that any known mass is produced by similar circumstances, on various occasions; such mass or object must necessarily contain all its qualities, and be equal to exhibit all its effects in hitherto untried events. Upon any occasion where we are either certain, or have a high probability, that an object presented to us is truly similar to a former one, and was created by the same causes; we expect all tried qualities to be the same as before, and any *untried* quality, (that is, any quality not in present operation, though previously ascertained,) must belong ever after to its definition. All that is necessary is to be correct, as to the prevening or influencing circumstances which gave *birth to the object.* They being the same on any two or more occasions, the object elicited must necessarily be the same – but it is not the same, unless it hath all its *qualities,* and no other than its qualities. Therefore fire, in order to have a right to the sign of the word fire, for an expression of its attributes, in order to be a "*like cause,*" must of necessity burn as much as it must be red, otherwise the red object were not fire; and *could not have been produced by those causes that elicit that element.* I mean therefore to conclude, that Effects are but the qualities of an object not experienced by some of the senses of the human frame, whilst certain others at present touch it; *the knowledge of which last,* being joined to the observation of the WHENCE the object was produced, beget the knowledge of what new untried qualities may be expected in future under given circumstances. It becomes therefore part of the definition of fire to burn certain bodies, to melt others; of bread to nourish the human body; of snow to be cold, and white; and these qualities they must have, in order to compose that entire *enumeration of qualities,* for which appropriate names have been formed, and to the exhibition of which similar and efficient causes have been in action.

If it should be said, that in considering objects as masses of combined qualities, the *result* of like Causes previously in action, we beg the question not yet supposed to be granted, – I answer; that *like* Causes, that is, *like* objects, are by the supposition admitted, and then the question arises, whether it is demonstrable they must have like effects or qualities, *under like circumstances in future?* I answer, they must have like effects, or qualities, because there is nothing else *given* that can be supposed to make a difference; and a difference of qualities could not arise of *itself,* could not begin its own existence; and I add, not only, there is nothing else *supposed* that can make a difference; but that when we also know that in the FORMATION of any object no difference took place; then, *there is no ground whatever,* for imagining the *possibility* of an *alteration* in *the effects of that object.* But although it be very difficult in the analysis of this question, not to use the word *cause* in its intended sense, before the definition of the word is given, and although it be true that in this last observation I may have done so in saying, *that objects must be the same which are elicited from like causes,* i.e. *from the junction of like prevening circumstances*; (and which position will be fully borne out in the process of the argument;) yet a fastidious reader may omit every such reference to the notion of Cause; for the argument is perfect without it, and stands thus:

Effects are nothing but those *same conjunctions of qualities,* which in other words are admitted as *similar causes,* in the supposition of the question. The objects (whose *union is necessary* to a given result,) must certainly exist, *antecedent* to such an union. But it is *in their* UNION, there exists those *newly formed objects,* or masses of qualities called *Effects,* which are therefore *identical* with the *similar cause;* for in *this union,* Cause and Effect are *synchronous,* and they are but different words for the same *Essence.* Fire and wood must be antecedent to combustion, no doubt; but in the *union of Fire and Wood,* there exists immediately *combustion* as a new event in nature; – also in this union exists the similar *cause* allowed by the *data,* whilst combustion is also termed the *Effect* of the union of Fire and Wood; but, however termed, an *effect,* is in fact a new but similar object as heretofore. A *similar* mass of qualities, in kind, which cannot therefore be a *different* mass of qualities in kind. – Equals added to equals upon any two occasions, the whole must be equal; add equal qualities to equal qualities, the sum of the qualities must be equal upon every repetition of the junction; – and the *sum* must be the *same result* taken *twice over,* not two *different,* or possibly *altered sums.* Therefore I repeat, that in the consideration of the nature of Cause and Effect, it is immaterial whether the yet unframed qualities of objects, previous to their junction, be named *effects;* they are to be considered as qualities; and qualities may be considered as *effects,* under any circumstances that prevent their usual exhibition. Effects when developed are no more than qualities; and qualities previous to their developement are in our imagination considered as Effects.

3dly. Again, it is immaterial to the definition of the relation of Cause and Effect, that we are not acquainted with the "secret powers" of natural objects, either before or after experience; for when we find, that in any distinct and given circumstances they put on certain qualities to the senses, their secret powers and properties must be qualified in all *like circumstances* to be the same, and are obliged to be so; because no contrary qualities could "begin their existences of themselves;" and by the *supposition* there is no *cause in the circumstances,* to give rise to any *differences in the qualities.* Indeed, Mr. Hume makes a great mistake in supposing it necessary to demonstrate, in every particular instance, what *particular* Effect must necessarily flow from its object, in order to gain the idea of *necessary Connexion.* The *how* and the *why* have nothing to do with the general reasoning affecting the general proposition; for "whether like Causes shall produce like Effects" is *not* a question exactly the same as whether "such particular causes shall have such *particular* effects? which Mr. Hume seems to consider as precisely of the same import[4]; whereas *one is a general* question, which however answered, in the *affirmative* or *negative,* would apply to *particulars.* But supposing in each *particular* instance under our notice, we could descry the "*secret powers of nature,*" the general question concerning *all* like causes would still remain unanswered; and an universal conclusion could not logically be deduced from the particular premises concerning it: as will be more fully argued in the discussion upon Dr. Brown's reasoning.

If it should be asked, (as Mr. Hume presently does,) how is it known when objects are similar upon any two occasions; the "sensible qualities may be the same, and not the *secret powers, upon which the Effects depend?*" I answer, this is to *shift* the question from the examination of *like Causes supposed,* to the consideration of the *method whereby their presence* may be *detected.*[5] But this difficulty is met, and considered in, its proper place; I shall only here say, that as the *secret powers* are the real external unknown Causes in Nature, which determine the sensible qualities, as well as *every other Effect*; so when we find the sensible qualities the same on any two occasions, we are sure the secret powers are similar *thus far,* and therefore fitted to exhibit their further similar effects; – (or *combined secret powers and sensible qualities;*) and although some *unobserved* cause might creep in to alter the object, whilst appearing the same, yet this we do not *imagine* when we are not aware of it, especially in cases where the same sensible qualities have been regularly exhibited along with like secret powers; for this regularity is perceived as an *Effect,* for which there must be a proportional *Cause,* and begets a proportional belief accordingly. – We argue from the *regular Effects,* (the sensible qualities;) to the *regular Causes,* (the secret powers;) which having been equal to certain other *Effects or properties,* we expect again the same, under similar circumstances. – We argue from the regular *ends* nature keeps in view, up to nature's God, who ordained them, and who must be supposed still to continue true to those ends; and along with the grander operations of nature, we may often in many cases observe our own actions, and those of others, conspiring *only* to fashion similar objects. But when the *secret powers,* and sensible qualities, are known, *or supposed the same,* the conclusion is demonstrative: *so must be the Effects.* Whilst, were it possible to know the secret powers in each *particular* past instance, *universal* truth would not thence result. Neither has Mr. Hume any right to make this argument; because to conceive "there may be secret powers which may change the Effects, dependent on them," is to make use of the relation between Cause and Effect, as of a really necessary connexion, in order to oppose his adversary: a principle which he previously refuses to admit. Also the objection forms an illogical argument in another way. For it virtually draws a general conclusion from *two negative premises.* To assert, that like sensible qualities merely, will NOT produce like Effects; and, that *like sensible qualities* are NOT *like Causes,* is to separate the middle term both from the subject and from the predicate of the general question. By *such an argument* Mr. Hume is certainly right in supposing, that REASON cannot support "*our conclusions concerning the operations of Cause and Effect.*"

Having thus cleared a way, towards the comprehension of this relation of Cause and Effect, we will proceed to a definition of those terms in the next Section.

Notes

1 Essays, Sec. 4. p. 27.

2 I conceive it impossible to have a complete conviction that every Effect is inherent, or contained in its Cause, until the mind be imbued with the knowledge, that objects are but unknown circumstances in Nature, when unperceived by the senses; which when perceived, exhibit their appropriate qualities accordingly; and which then appear in certain defined masses, as to the different senses they affect, as to their figure, &c.; and receive an arbitrary name for their assemblage. They must have also among each other certain proportions. When these unknown circumstances, (or affections, or substances,) in nature, *mix,* and are thereby *altered, the qualities which affect the senses* are in the *same proportions* altered, and are necessarily included in those objects as their Effects. But this part of the subject, is of such moment that a separate consideration of it is intended.

3 When more trials are needed than ONE, it is in order to *detect* the circumstances, not to lay *a foundation for the general principle,* that a LIKE Cause repeated, a LIKE Effect will take place.

4 Compare Sec. 4. p. 30, with Sec. 4. p. 34.

5 I should not here have taken notice of this objection, but that as Mr. Hume does suddenly shift the question, so I would not appear to avoid an answer to it: otherwise it is something too early to enter upon the subject; obliging me to make use of my argument previously to its complete developement. But the reader may pass over to the next Section if he please.

Part 2

SIR JOHN F. W. HERSCHEL'S *PRELIMINARY DISCOURSE*

Editorial Headnote

7. John F. W. Herschel, *A Preliminary Discourse on the Study of Natural Philosophy* (1830)

8. William Whewell, "[Review of] *A Preliminary Discourse on the Study of Natural Philosophy,*" *The Quarterly Review* 45:90 (1831)

By publication date, the first of the major nineteenth-century British works on scientific methodology is the *Preliminary Discourse* of Sir John F. W. Herschel (1792–1871). Herschel was the son of the renowned astronomer William Herschel (1738–1822), discoverer of Uranus, and, after some hesitation, threw himself fully into a scientific career, becoming immensely influential in astronomy, chemistry, and the invention of photography, as well as important to organizational activity at both the Royal Society and the British Association for the Advancement of Science.

Herschel's book was published in a series of inexpensive texts for wide public distribution, the "Cabinet Cyclopædia." As Jim Secord has argued, this makes it almost certain that he targeted both an audience of his fellow scientists and a general public that saw in the scientific method a means by which they could improve their character.[1] The genre of "treatise in general philosophy of science," after all, was in its infancy. We see this from the very outset of the work, where Herschel discusses the personal virtues that are required to understand the importance and utility of theoretical work in the sciences – presented first and foremost as the inner characteristics and habits of mind, both rigorous but also joyful and a source of pleasure, that a good scientist must possess. Considering those habits could be a source of refinement for people from any walk of life.

The bulk of the excerpt concerns Herschel's discussion of the proposal and evaluation of causal explanations for natural phenomena. Explanation, for Herschel, just is causal explanation: we explain when we have identified the "immediate producing cause" (§137). The process begins with the assertion that the advance of science has given us a kind of stock of known causes, the operation of which is already confirmed in nature (§138); Herschel, borrowing a phrase of Newton's from his rules for philosophizing, calls these *veræ causæ*, or "true causes." Such causes are, at the very least, known to operate in some way or another in the natural world. We might not have enough knowledge to be certain that a given cause has produced a particular effect in which we are interested, but we know that it is at least "a cause on which a philosopher may consent to reason" (§139). Herschel cites approvingly the geology of Lyell, to which we will return in Volume 2, as proposing causes like subsidence and erosion – the power of which we cannot deny – as the motor for large-scale, long-term geological changes. Those causes can't be immediately seen as responsible for something like the formation of mountain ranges, but they give us nonetheless a point of departure.

It is here that Herschel makes reference to a feature that has often been cited as distinctive of his methodological approach: the use of analogy (§142). The better we are at building analogies, the better we will be at recognizing when some well-known vera causa *might* have given rise to a phenomenon that we hope to explain. Experiment can be relied upon to help us in cases where we aren't in possession of the right kinds of phenomena to distinguish between two or more potential causes; this is Herschel's interpretation of Bacon's "crucial experiments" (§144). Anticipating the development of the rules of thumb now known as Mill's methods (to which we will return later in this volume), Herschel builds a theory of our search for the "common cause" of some "great mass of assembled facts" (§§145–62).

But this is only our first step: having arrived at a potential *vera causa*, we must now verify that it was actually the cause responsible for the phenomenon we are explaining (§§170–75). This first amounts to checking every known instance of the phenomenon to make sure that none have escaped our attention, taking careful account of any potential exceptions. In particular, exceptions might show us the existence of what Herschel calls "residual phenomena," or leftover features, unexplained by the set of all of our known causal explanations, which could lead us to further discoveries (§174). Finally, anticipating Whewell on consilience (see the next selection), we should consider "cases not originally contemplated" in elaborating our novel cause, or even of "that very kind which were at first considered hostile to them" (§§176–83).

When we arrive at the limits of human knowledge, however, we will be forced to generate speculative hypotheses; no known causal forces can explain, for instance, the ways in which heat or light function, Herschel argues (§207). He does not wish to bar the generation of such theories – they might, and in the case of the wave theory of light, he thinks they have – become so well confirmed that they at least "run so close a parallel" with the truth that they serve as incredibly fruitful generators of future experiment and invention (§208).

Herschel's work was immediately appreciated by a wide swath of the British scientific community,[2] as can be seen from the extensive review of it by William Whewell (1794–1866) that was published in the *Quarterly Review*. Whewell here not only describes and elaborates on Herschel's book, but starts to sketch the differences between the approaches of the two men, differences which will become even clearer in the next selection from Whewell's own masterwork. Noting with approval that it is "one of the first considerable attempts to expound in any detail the rules and doctrines of" contemporary science (p. 99), he devotes much space to considering the sense in which Herschel's work might be said to be "Baconian."[3] Reid had defined "induction" as our common-sense belief in the ongoing uniformity of the laws of nature; for Whewell, and by implication for Herschel as well, it is "the process of considering a class of phenomena, or two associated classes of phenomena, as represented by a general *law*, or single conception of the mind" (p. 100). As Lukas Verburgt has argued, for authors like Herschel and Whewell both, "to make room for hypothesizing . . . was not to dismiss Bacon but rather to formulate a Baconianism philosophically attuned to the historical record."[4]

Whewell's primary and extensively illustrated example, which Herschel also discussed in portions of the *Preliminary Discourse* not excerpted here, is the theory of heat. This theory offers a paradigmatic case of a move between creating novel inductive justifications – proposing and evaluating hypotheses – and drawing out their consequences for further testing and unification. But his hesitation about speculative theories of heat (caloric, or the theory of heat as a kind of fluid, in this case) presages his broader worry about Herschel's work: that Herschel does not suitably protect us against the possibility that we could be carried away by our hypotheses, "anticipating" a law rather than proceeding via "general induction" (p. 111). "It is easy," Whewell writes, "to form such conjectures, and hardly more meritorious to be right than wrong in them" (p. 113). Whewell closes by emphasizing Herschel's view of science as morally and mentally uplifting, even chiding Herschel for having spoken too highly of the application of scientific knowledge to everyday life, rather than the abstract pursuit of truth.

Notes

1 J. A. Secord, *Visions of Science: Books and Readers at the Dawn of the Victorian Age* (Chicago: University of Chicago Press, 2014), pp. 80–106.
2 For more, see C. H. Pence, "John Herschel's Methodology in the Scientific Community," in S. Case and L. M. Verburgt (eds), *The Cambridge Companion to John Herschel* (Cambridge: Cambridge University Press, 2024), pp. 257–79.
3 A preoccupation which deeply concerned both men, as we know from L. J. Snyder, *The Philosophical Breakfast Club* (New York: Broadway, 2011).
4 L. M. Verburgt, 'Scientific Method, Induction, and Probability: The Whewell-de Morgan Debate on Baconianism, 1830s–1850s,' *HOPOS* 14:1 (2024), pp. 134–63, on p. 140.

7

PRELIMINARY DISCOURSE ON THE STUDY OF NATURAL PHILOSOPHY

John F. W. Herschel

Source: John F. W. Herschel, *Preliminary Discourse on the Study of Natural Philosophy*, 2nd edn (1851), Chapter I, pp. 13–17; Chapter V, pp. 135–138; Chapter VI, pp. 144–175; Chapter VII, pp. 190–200.

(9.) After all, however, it must be confessed, that to minds unacquainted with science, and unused to consider the mutual dependencies of its various branches, there is something neither unnatural nor altogether blamable in the ready occurrence of this question of direct advantage. It requires some habit of abstraction, some penetration of the mind with a tincture of scientific enquiry, some conviction of the value of those estimable and treasured principles which lie concealed in the most common and homely facts, – some experience, in fine, of success in developing and placing them in evidence, announcing them in precise terms, and applying them to the explanation of other facts of a less familiar character, or to the accomplishment of some obviously useful purpose: – to cure the mind of this tendency to rush at once upon its object, to undervalue the means in over-estimation of the end, and while gazing too intently at the goal which alone it has been accustomed to desire, to lose sight of the richness and variety of the prospects that offer themselves on either hand on the road.

(10.) We must never forget that it is principles, not phenomena, – the interpretation, not the mere knowledge of facts, – which are the objects of enquiry to the natural philosopher. As truth is single, and consistent with itself, a principle may be as completely and as plainly elucidated by the most familiar and simple fact, as by the most imposing and uncommon phenomenon. The colours which glitter on a soap-bubble are the immediate consequence of a principle the most important from the variety of phenomena it explains, and the most beautiful, from its simplicity and compendious neatness, in the whole science of optics. If the nature of periodical colours can be made intelligible by the contemplation of such a trivial object, from that moment it becomes a noble instrument in the eye of correct judgment; and to blow a large, regular, and durable soap-bubble may become the serious and praiseworthy endeavour of a sage, while children stand round and scoff, or children of a larger growth hold up their hands in astonishment at such waste of time and trouble. To the natural philosopher there is no natural object unimportant or trifling. From the

DOI: 10.4324/9781003263814-9

least of nature's works he may learn the greatest lessons. The fall of an apple to the ground may raise his thoughts to the laws which govern the revolutions of the planets in their orbits; or the situation of a pebble may afford him evidence of the state of the globe he inhabits, myriads of ages ago, before his species became its denizens.

(11.) And this is, in fact, one of the great sources of delight which the study of natural science imparts to its votaries. A mind which has once imbibed a taste for scientific enquiry, and has learnt the habit of applying its principles readily to the cases which occur, has within itself an inexhaustible source of pure and exciting contemplations: – one would think that Shakspeare had such a mind in view when he describes a contemplative man as finding all nature eloquent – the very trees, the brooks, and the stones reading to him lessons of deep and serious import. Accustomed to trace the operation of general causes, and the exemplification of general laws, in circumstances where the uninformed and unenquiring eye perceives neither novelty nor beauty, he walks in the midst of wonders: every object which falls in his way elucidates some principle, affords some instruction, and impresses him with a sense of harmony and order. Nor is it a mere passive pleasure which is thus communicated. A thousand questions are continually arising in his mind, a thousand subjects of enquiry presenting themselves, which keep his faculties in constant exercise, and his thoughts perpetually on the wing, so that lassitude is excluded from his life, and that craving after artificial excitement and dissipation of mind, which leads so many into frivolous, unworthy, and destructive pursuits, is altogether eradicated from his bosom.

(12.) It is not one of the least advantages of these pursuits, which, however, they possess in common with every class of intellectual pleasures, that they are altogether independent of external circumstances, and are to be enjoyed in every situation in which a man can be placed in life. The highest degrees of worldly prosperity are so far from being incompatible with them, that they supply inestimable advantages for their pursuit, and that sort of fresh and renewed relish which arises partly from the sense of contrast, partly from experience of the peculiar preeminence they possess over the pleasures of sense in their capability of unlimited increase and continual repetition without satiety or distaste. They may be enjoyed, too, in the intervals of the most active business; and the calm and dispassionate interest with which they fill the mind renders them a most delightful retreat from the agitations and dissensions of the world, and from the conflict of passions, prejudices, and interests in which the man of business finds himself involved. There is something in the contemplation of general laws which powerfully induces and persuades us to merge individual feeling, and to commit ourselves unreservedly to their disposal; while the observation of the calm, energetic regularity of nature, the immense scale of her operations, and the certainty with which her ends are attained, tends, irresistibly, to tranquillize and re-assure the mind, and render it less accessible to repining, selfish, and turbulent emotions. And this it does, not by debasing our nature into weak compliances and abject submission to circumstances, but by filling us, as from an inward spring, with a sense of nobleness and power which enables us to rise superior to them; by showing us our strength and innate dignity, and by calling upon us for the exercise of those powers and faculties by which we are susceptible of the comprehension of so much greatness, and which form, as it were, a link between ourselves and the best and noblest benefactors of our species, with whom we hold communion in thoughts and participate in discoveries which have raised them above their fellow-mortals, and brought them nearer to their Creator.

CHAP. V.

OF THE CLASSIFICATION OF NATURAL OBJECTS AND PHENOMENA, AND OF NOMENCLATURE.

(129.) THE number and variety of objects and relations which the observation of nature brings before us are so great as to distract the attention, unless assisted and methodized by such judicious distribution of them in classes as shall limit our view to a few at a time, or to groups so bound together by general resemblances that, for the immediate purpose for which we consider them, they may be regarded as individuals. Before we can enter into any thing which deserves to be called a general and systematic view of nature, it is necessary that we should possess an enumeration, if not complete, at least of considerable extent, of her materials and combinations; and that those which appear in any degree important should be distinguished by names which may not only tend to fix them in our recollection, but may constitute, as it were, nuclei or centres, about which information may collect into masses. The imposition of a name on any subject of contemplation, be it a material object, a phenomenon of nature, or a group of facts and relations, looked upon in a peculiar point of view, is an epoch in its history of great importance. It not only enables us readily to refer to it in conversation or writing, without circumlocution, but, what is of more consequence, it gives it a recognized existence in our own minds, as a matter for separate and peculiar consideration; places it on a list for examination; and renders it a head or title, under which information of various descriptions may be arranged; and, in consequence, fits it to perform the office of a connecting link between all the subjects to which such information may refer.

(130.) For these purposes, however, a temporary or provisional name, or one adapted for common parlance, may suffice. But when a very great multitude of objects come to be referred to one class, especially of such as do not offer very obvious and remarkable distinctions, a more systematic and regular nomenclature becomes necessary, in which the names shall recall the differences as well as the resemblances between the individuals of a class, and in which the direct relation between the name and the object shall materially assist the solution of the problem, "*given the one, to determine the other.*" How necessary this may become, will be at once seen, when we consider the immense number of individual objects, or rather species, presented by almost every branch of science of any extent; which absolutely require to be distinguished by names. Thus, the botanist is conversant with from

80,000 to 100,000 species of plants; the entomologist with, perhaps, as many, of insects: the chemist has to register the properties of combinations, by twos, threes, fours, and upwards, in various doses of upwards of fifty different elements, all distinguished from each other by essential differences; and of which though a great many thousands are known, by far the greater part have never yet been formed, although hundreds of new ones are coming to light, in perpetual succession, as the science advances; all of which are to be named as they arise. The objects of astronomy are, literally, as numerous as the stars of heaven; and although not more than one or two thousand require to be expressed by distinct names, yet the number, respecting which particular information is required, is not less than a hundred times that amount; and all these must be registered in lists, (so as to be at once referred to, and so that none shall escape,) if not by actual names, at least by some equivalent means.

(131.) Nomenclature, then, is, in itself, undoubtedly an important part of science, as it prevents our being lost in a wilderness of particulars, and involved in inextricable confusion. Happily, in those great branches of science where the objects of classification are most numerous, and the necessity for a clear and convenient nomenclature most pressing, no very great difficulty in its establishment is felt. The very multitude of the objects themselves affords the power of grouping them in subordinate classes, sufficiently well defined to admit of names, and these again into others, whose names may become attached to, or compounded with, the former, till at length the particular species is identified. The facility with which the botanist, the entomologist, or the chemist, refers by name to any individual object in his science shows what may be accomplished in this way when characters are themselves distinct. In other branches, however, considerable difficulty is experienced. This arises mostly where the species to be distinguished are separated from each other chiefly by difference in degree, of certain qualities common to all, and where the degrees shade into each other insensibly. Perhaps such subjects can hardly be considered ripe for systematic nomenclature; and that the attempt to apply it ought only to be partial, embracing such groups and parcels of individuals as agree in characters evidently natural and generic, and leaving the remainder under trivial or provisional denominations, till they shall be better known, and capable of being scientifically grouped.

[. . .]

CHAP. VI.

OF THE FIRST STAGE OF INDUCTION. – THE DISCOVERY OF PROXIMATE CAUSES, AND LAWS OF THE LOWEST DEGREE OF GENERALITY, AND THEIR VERIFICATION.

(137.) The first thing that a philosophic mind considers, when any new phenomenon presents itself, is its *explanation*, or reference to an immediate producing cause. If that cannot be ascertained, the next is to *generalize* the phenomenon, and include it, with others analogous to it, in the expression of some law, in the hope that its consideration, in a more advanced state of knowledge, may lead to the discovery of an adequate proximate cause.

(138.) Experience having shown us the manner in which one phenomenon depends on another in a great variety of cases, we find ourselves provided, as science extends, with a continually increasing stock of such antecedent phenomena, or causes (meaning at present merely proximate causes), competent, under different modifications, to the production of a great multitude of effects, besides those which originally led to a knowledge of them. To such causes Newton has applied the term *veræ causæ;* that is, causes recognized as having a real existence in nature, and not being mere hypotheses or figments of the mind. To exemplify the distinction: – The phenomenon of shells found in rocks, at a great height above the sea, has been attributed to several causes. By some it has been ascribed to a plastic virtue in the soil; by some, to fermentation; by some, to the influence of the celestial bodies; by some, to the casual passage of pilgrims with their scallops; by some, to birds feeding on shell-fish; and by all modern geologists, with one consent, to the life and death of real mollusca at the bottom of the sea, and a subsequent alteration of the relative level of the land and sea. Of these, the plastic virtue and celestial influence belong to the class of figments of fancy. Casual transport by pilgrims is a real cause, and might account for a few shells here and there dropped on frequented passes, but is not extensive enough for the purpose of explanation. Fermentation, generally, is a real cause, so far as that there *is such a thing;* but it is not a real cause of the production of a shell in a rock, since no such thing was ever witnessed as one of its effects, and rocks and stones do not ferment. On the other hand, for a shell-fish dying at the bottom of the sea to leave his shell in the mud, where it becomes silted over and imbedded, happens daily; and the elevation of the bottom of the sea to become dry land has really been witnessed so often, and on such a scale, as to qualify it for a *vera causa* available in sound philosophy.

(139.) To take another instance, likewise drawn from the same deservedly popular science: – The fact of a great change in the general climate of large tracts of the globe, if not of the whole earth, and of a diminution of general temperature, having been recognised by geologists, from their examination of the remains of animals and vegetables of former ages enclosed in the strata, various causes for such diminution of temperature have been assigned. Some consider the whole globe as having gradually cooled from absolute fusion; some regard the immensely superior activity of former volcanoes, and consequent more copious communication of internal heat to the surface, in former ages, as the cause. Neither of these can be regarded as real causes in the sense here intended; for we do not *know* that the globe has so cooled from fusion, nor are we sure that such supposed greater activity of former than of present volcanoes really did exist. A cause, possessing the essential requisites of a *vera causa,* has, however, been brought forward[1] in the varying influence of the distribution of land and sea over the surface of the globe: a change of such distribution, in the lapse of ages, by the degradation of the old continents, and the elevation of new, being a demonstrated fact; and the influence of such a change on the climates of particular regions, if not of the whole globe, being a perfectly fair conclusion, from what we know of continental, insular, and oceanic climates by actual observation. Here, then, we have, at least, a cause on which a philosopher may consent to reason; though, whether the changes actually going on are such as to warrant the whole extent of the conclusion, or are even taking place in the right direction, may be considered as undecided till the matter has been more thoroughly examined.

(140.) To this we may add another, which has likewise the essential characters of a *vera causa* in the astronomical *fact* of the actual slow diminution of the eccentricity of the earth's orbit round the sun; and which, as a general one, affecting the *mean temperature of the whole globe,* and as one of which the effect is both inevitable, and susceptible, to a certain degree, of exact estimation, deserves consideration. It is evident that the *mean* temperature of the whole surface of the globe, in so far as it is maintained by the action of the sun at a higher degree than it would have were the sun extinguished, must depend on the mean quantity of the sun's rays which it receives, or, which comes to the same thing, on the *total* quantity received in a given invariable time: and the length of the year being unchangeable in all the fluctuations of the planetary system, it follows, that the total *annual* amount of solar radiation will determine, *cæteris paribus*, the general climate of the earth. Now, it is not difficult to show that this amount is inversely proportional to the minor axis of the ellipse described by the earth about the sun, regarded as slowly variable; and that, therefore, the major axis remaining, as we know it to be, constant, and the orbit being actually in a state of approach to a circle, and, consequently, the minor axis being on the *increase*, the mean annual amount of solar radiation received by the whole earth must be actually on the *decrease*. We have here, therefore, an evident real cause, of sufficient universality, and acting *in the right direction*, to account for the phenomenon. Its adequacy is another consideration.[2]

(141.) Whenever, therefore, any phenomenon presents itself for explanation, we naturally seek, in the first instance, to refer it to some one or other of those real causes which experience has shown to exist, and to be efficacious in producing similar phenomena. In this attempt our probability of success will, of course, mainly depend, 1st, On the number and variety of causes experience has placed at our disposal; 2dly, On our habit of applying them to the explanation of natural phenomena; and, 3dly, On the number of analogous phenomena we can collect, which have either been explained, or which admit of explanation by some one or other of those causes, and the closeness of their analogy with that in question.

(142.) Here, then, we see the great importance of possessing a stock of analogous instances or phenomena which class themselves with that under consideration, the explanation of one among which may naturally be expected to lead to that of all the rest. If the analogy of two phenomena be very close and striking, while, at the same time, the cause of one is very obvious, it becomes scarcely possible to refuse to admit the action of an analogous cause in the other, though not so obvious in itself. For instance, when we see a stone whirled round in a sling, describing a circular orbit round the hand, keeping the string stretched, and flying away the moment it breaks, we never hesitate to regard it as retained in its orbit by the tension of the string, that is, by *a force* directed to the centre; for we feel that we do really exert such a force. We have here *the direct perception* of the cause. When, therefore, we see a great body like the moon circulating round the earth and not flying off, we cannot help believing it to be prevented from so doing, not indeed by a material tie, but by that which operates in the other case through the intermedium of the string, – a *force* directed constantly to the centre. It is thus that we are continually acquiring a knowledge of the existence of causes acting under circumstances of such concealment as effectually to prevent their direct discovery.

(143.) In general we must observe that motion, wherever produced or changed, invariably points out the existence of *force* as its cause; and thus the forces of nature become known and measured by the motions they produce. Thus, the *force* of magnetism becomes known by the deviation produced by iron in a compass needle, or by a needle leaping up to a magnet held over it, as certainly as by that adhesion to it, when in contact and at rest, which requires force to break the connection; and thus the currents produced in the surface of a quantity of quicksilver, electrified under a conducting fluid, have pointed out the existence and direction of forces of enormous intensity developed by the electric circuit, of which we should not otherwise have had the least suspicion.[3]

(144.) But when the cause of a phenomenon neither presents itself obviously on the consideration of the phenomenon itself, nor is as it were forced on our attention by a case of strong analogy, such as above described, we have then no resource but in a deliberate assemblage of all the parallel instances we can muster; that is, to the formation of a class of facts, having the phenomenon in question for a head of classification; and to a search among the individuals of this class for some other common points of agreement, among which the cause will of necessity be found. But if more than one cause should appear, we must then endeavour to find, or, if we cannot find, to *produce*, *new facts*, in which each of these in succession shall be wanting, while yet they agree in the general point in question. Here we find the use of what Bacon terms *"crucial instances,"* which are phenomena brought forward to decide between two causes, each having the same analogies in its favour. And here, too, we perceive the utility of *experiment* as distinguished from mere passive observation. We make an experiment of the crucial kind when we form combinations, and put in action causes from which some particular one shall be deliberately excluded, and some other purposely admitted; and by the agreement or disagreement of the resulting phenomena with those of the class under examination, we decide our judgment.

(145.) When we would lay down general rules for guiding and facilitating our search, among a great mass of assembled facts, for their common cause, we must have regard to the characters of that relation which we intend by cause and effect. Now, these are, –

1st, Invariable connection, and, in particular, invariable antecedence of the cause and consequence of the effect, unless prevented by some counteracting cause. But it must be observed, that, in a great number of natural phenomena, the effect is produced gradually, while the cause often goes on increasing in intensity; so that the antecedence of the one and consequence of the other becomes difficult to trace, though it really exists. On the other hand, the effect often follows the cause so instantaneously, that the interval cannot be perceived. In consequence of this, it is sometimes difficult to decide, of two phenomena constantly accompanying one another, which is cause or which effect.

2d, Invariable negation of the effect with absence of the cause, unless some other cause be capable of producing the same effect.

3d, Increase or diminution of the effect, with the increased or diminished intensity of the cause, in cases which admit of increase and diminution.

4th, Proportionality of the effect to its cause in all cases of *direct unimpeded* action.

5th, Reversal of the effect with that of the cause.

(146.) From these characters we are led to the following observations, which may be considered as so many propositions readily applicable to particular cases, or rules of philosophizing: we conclude, 1st, That if in our group of facts there be one in which any assigned peculiarity, or attendant circumstance, is wanting or opposite, such peculiarity cannot be the cause we seek.

(147.) 2d, That any circumstance in which all the facts without exception agree, *may* be the cause in question, or, if not, at least a collateral effect of the same cause: if there be but one such point of agreement, this possibility becomes a certainty; and, on the other hand, if there be more than one, they may be concurrent causes.

(148.) 3d, That we are not to deny the existence of a cause in favour of which we have a unanimous agreement of strong analogies, though it may not be apparent how such a cause can produce the effect, or even though it may be difficult to conceive its existence under the circumstances of the case; in such cases we should rather appeal to experience when possible, than decide *à priori* against the cause, and try whether it cannot be made apparent.

(149.) For instance: seeing the sun vividly luminous, every analogy leads us to conclude it intensely hot. How heat can produce light, we know not; and how such a heat can be maintained, we can form no conception. Yet we are not, therefore, entitled to deny the inference.

(150.) 4th, That contrary or opposing facts are equally instructive for the discovery of causes with favourable ones.

(151.) For instance: when air is confined with moistened iron filings in a close vessel over water, its bulk is diminished, by a certain portion of it being abstracted and combining with the iron, producing *rust*. And, if the remainder be examined, it is found that it will *not* support flame or animal life. This contrary fact shows that the cause of the support of flame and animal life is to be looked for in that part of the air which the iron abstracts, and which rusts it.

(152.) 5th, That causes will very frequently become obvious, by a mere arrangement of our facts in the order of intensity in which some peculiar quality subsists; though not of necessity, because counteracting or modifying causes may be at the same time in action.

(153.) For example: sound consists in impulses communicated to our ears by the air. If a series of impulses of equal force be communicated to it at equal intervals of time, at first in slow succession, and by degrees more and more rapidly, we hear at first a rattling noise, then a low murmur, and then a hum, which by degrees acquires the character of a musical note, rising higher and higher in acuteness, till its pitch becomes too high for the ear to follow. And from this correspondence between the pitch of the note and the rapidity of succession of the impulse, we conclude that our sensation of the different pitches of musical notes originates in the different rapidities with which their impulses are communicated to our ears.

(154.) 6th, That such counteracting or modifying causes may subsist unperceived, and annul the effects of the cause we seek, in instances which, but for their action, would have come into our class of favourable facts; and that, therefore, exceptions may often be made to disappear by removing or allowing for such counteracting causes. This remark becomes of the greatest importance, when (as is often the case) a single striking exception stands out, as it were, against an otherwise unanimous array of facts in favour of a certain cause.

(155.) Thus, in chemistry, the *alkaline* quality of the alkaline and earthy bases is found to be due to the presence of oxygen combined with one or other of a peculiar set of metals. Ammonia is, however, a violent outstanding exception, such as here alluded to, being a compound of azote and hydrogen: but there are almost certain indications that this exception is not a real one, but assumes that appearance in consequence of some modifying cause not understood.

(156.) 7th, If we can either find produced by nature, or produce designedly for ourselves, two instances which agree *exactly* in all but one particular, and differ in that one, its influence in producing the phenomenon, if it have any, *must* thereby be rendered sensible. If that particular be present in one instance and wanting altogether in the other, the production or non-production of the phenomenon will decide whether it be or be not the only cause: still more evidently, if it be present *contrariwise* in the two cases, and the effect be thereby reversed. But if its total presence or absence only produces a change in the *degree* or intensity of the phenomenon, we can then only conclude that it acts as a concurrent cause or condition with some other to be sought elsewhere. In nature, it is comparatively rare to find instances pointedly differing in one circumstance and agreeing in every other; but when we call experiment to our aid, it is easy to produce them; and this is, in fact, the grand application of *experiments of enquiry* in physical researches. They become more valuable, and their results clearer, in proportion as they possess this quality (of agreeing exactly in all their circumstances but one), since the question put to nature becomes thereby more pointed, and its an swer more decisive.

(157.) 8th, If we cannot obtain a complete negative or opposition of the circumstance whose influence we would ascertain, we must endeavour to find cases where it varies considerably in degree. If *this* cannot be done, we may perhaps be able to weaken or exalt its influence by the introduction of some fresh circumstance, which, abstractedly considered, seems *likely* to produce this effect, and thus obtain indirect evidence of its influence. But then we are always to remember, that the evidence so obtained *is* indirect, and that the new circumstance introduced *may* have a direct influence of its own, or may exercise a modifying one on some *other* circumstance.

(158.) 9th, Complicated phenomena, in which several causes concurring, opposing, or quite independent of each other, operate at once, so as to produce a compound effect, may be simplified by subducting the effect of all the known causes, as well as the nature of the case permits, either by deductive reasoning or by appeal to experience, and thus leaving, as it were, a *residual phenomenon* to be explained. It is by this process, in fact, that science, in its present advanced state, is chiefly promoted. Most of the phenomena which nature presents are very complicated; and when the effects of all known causes are estimated with exactness, and subducted, the residual facts are constantly appearing in the form of phenomena altogether new, and leading to the most important conclusions.

(159.) For example: the return of the comet predicted by professor Encke, a great many times in succession, and the general good agreement of its calculated with its observed place during any one of its periods of visibility, would lead us to say that its gravitation towards the sun and planets is the sole and sufficient cause of all the phenomena of its orbitual motion; but when the effect of this cause is strictly calculated and subducted from the observed motion, there is found to remain behind a *residual phenomenon,* which would never have been otherwise ascertained to exist, which is a small anticipation of the time of its reappearances or a diminution of its periodic time, which cannot be accounted for by gravity, and whose cause is therefore to be enquired into. Such an anticipation would be caused by the resistance of a medium dis-seminated through the celestial regions; and as there are other good reasons for believing this to be a *vera causa,* it has therefore been ascribed to such a resistance.

(160.) This 9th observation is of such importance in science, that we shall exemplify it by another instance or two. M. Arago, having suspended a magnetic needle by a silk thread, and set it in vibration, observed, that it came much sooner to a state of rest when suspended over a plate of copper, than when no such plate was beneath it. Now, in both cases there were two *veræ causæ* why it *should* come at length to rest, viz. the resistance of the air, which opposes, and at length destroys, all motions performed in it; and the want of perfect mobility in the silk thread. But the effect of these causes being exactly known by the observation made in the absence of the copper, and being thus allowed for and subducted, a *residual* phenomenon appeared, in the fact that a retarding influence was exerted by the copper itself; and this fact, once ascertained, speedily led to the knowledge of an entirely new and unexpected class of relations. To add one more instance. If it be true (as M. Fourrier considers it demonstrated to be) that the celestial regions have a temperature independent of the sun, not greatly inferior to that at which quicksilver congeals, and much superior to some degrees of cold which have been artificially produced, two causes suggest themselves: one is that assigned by the author above mentioned; the radiation of the stars; another may be proposed in the ether or elastic medium mentioned in the last section, which the phenomena of light and the resistance of comets give us reason to believe fills all space, and which, in analogy to all the elastic media known, may be supposed to possess a temperature and a specific heat of its own, which it is capable of communicating to bodies surrounded by it. Now, if we consider that the heat radiated by the sun follows the same proportion as its light, and regard it as reasonable to admit with respect to stellar heat what holds good of solar; the effect of stellar radiation in maintaining a temperature in space should be as much inferior to that of the radiation of the sun as the light of a moonless midnight is to that of an equatorial noon; that is to say, almost inconceivably smaller. Allowing, then, the full effect for this cause, there would still remain a great residuum due to the presence of the ether.

(161.) Many of the new elements of chemistry have been detected in the investigation of *residual phenomena*. Thus, Arfwedson discovered lithia by perceiving an *excess of weight* in the sulphate produced from a small portion of what he considered as magnesia present in a mineral he had analysed. It is on this principle, too, that the *small concentrated residues of great operations* in the arts are almost sure to be the lurking places of new chemical ingredients: witness iodine, brome, selenium, and the new metals accompanying platina in the experiments of Wollaston and Tennant. It was a happy thought of Glauber to examine what every body else threw away.

(162.) Finally, we have to observe, that the detection of a *possible* cause, by the comparison of assembled cases, *must* lead to one of two things: either, 1st, The detection of a real cause, and of its manner of acting, so as to furnish a complete explanation of the facts; or, 2dly, The establishment of an abstract law of nature, pointing out two phenomena of a general kind as invariably connected; and asserting, that where one is, there the other will always be found. Such invariable connection is itself a phenomenon of a higher order than any particular fact; and when many such are discovered, we may again proceed to classify, combine, and examine them, with a view to the detection of *their* causes, or the discovery of still more general laws, and so on without end.

(163.) Let us now exemplify this inductive search for a cause by one general example: suppose *dew* were the phenomenon proposed, whose cause we would know. In the first place, we must separate dew from rain and the moisture of fogs, and limit the application of the term to what is really meant, which is, the spontaneous appearance of moisture on substances exposed in the open air when no rain or *visible* wet is falling. Now, here we have analogous phenomena in the moisture which bedews a cold metal or stone when we breathe upon it; that which appears on a glass of water fresh from the well in hot weather; that which appears on the *inside* of windows when sudden rain or hail chills the external air; that which runs down our walls when, after along frost, a warm moist thaw comes on: all these instances agree in one point (Rule 2. § 147.), the coldness of the object dewed, in comparison with the air in contact with it.

(164.) But, in the case of the night dew, is this a *real cause* – is it a fact that the object dewed *is* colder than the air? Certainly not, one would at first be inclined to say; for what is to *make* it so? But the analogies are cogent and unanimous; and, therefore, (pursuant to Rule 3. § 148.) we are not to discard their indications; and, besides, the experiment is easy: we have only to lay a thermometer in contact with the dewed substance, and hang one at a little distance above it out of reach of its influence. The experiment has been therefore made; the question has been asked, and the answer has been invariably in the *affirmative.* Whenever an object contracts dew, *it is* colder than the air. Here, then, we have *an invariable concomitant* circumstance: but is this chill an effect of dew, or its cause? That dews are accompanied with a chill is a common remark; but vulgar prejudice would make the cold the *effect* rather than the cause. We must, therefore, collect more facts, or, which comes to the same thing, vary the circumstances; since every instance in which the circumstances differ is a fresh fact; and, especially, we must note the contrary or negative cases (Rule 4. § 150.), *i.e.* where no dew is produced.

(165.) Now, 1st, no dew is produced on the surface of *polished metals*, but it *is* very copiously on glass, both exposed with their faces upwards, and in some cases the under side of a horizontal plate of glass is also dewed; which last circumstance (by Rule 1. § 146.) excludes the *fall* of moisture from the sky in an invisible form, which would naturally suggest itself as a cause. In the cases of polished metal and polished glass, the contrast shows evidently that the *substance* has much to do with the phenomenon; therefore, let the substance *alone* be diversified as much as possible, by exposing polished surfaces of various kinds. This done, *a scale of intensity* becomes obvious (Rule 5. § 152.). Those polished substances are found to be most strongly dewed which conduct heat worst; while those which conduct well resist dew most effectually. Here we encounter a *law* of the first degree of generality. But, if we expose rough surfaces, instead of polished, we sometimes find this law interfered with (Rule 5. § 152.). Thus, roughened iron, especially if painted over or blackened, becomes dewed sooner than varnished paper: the kind of *surface* therefore has a great influence. Expose, then, the *same* material in very diversified states as to surface (Rule 7. § 156.), and another scale of intensity becomes at once apparent; those *surfaces* which *part with their heat* most readily by radiation are found to contract dew most copiously: and thus we have detected another law of the same generality with the former, by a comparison of two classes of facts, one relating to dew, the other to the radiation of heat from surfaces. Again, the influence ascertained to exist of *substance* and *surface* leads us to consider that of *texture*: and here, again, we are presented on trial with remarkable differences, and with a third *scale of intensity*, pointing out substances of a close firm texture, such as stones, metals, &c. as unfavourable, but those of a loose one, as cloth, wool, velvet, eiderdown, cotton, &c. as eminently favourable, to the contraction of dew: and these are precisely those which are best adapted for clothing, or for impeding the free passage of heat from the skin into the air, so as to allow their outer surfaces to be very cold while they remain warm within.

(166.) Lastly, among the negative instances, (§ 150.) it is observed, that dew is never copiously deposited in situations much screened from the open sky, and not at all in *a cloudy night;* but if the clouds withdraw, even for a few minutes, and leave a clear opening, a deposition of dew presently begins, and goes on increasing. Here, then, a cause is distinctly pointed out by its antecedence to the effect in question (§ 145.). A clear view of the cloudless sky, then, is an essential condition, or, which comes to the same thing, clouds or surrounding objects act as *opposing causes.* This is so much the case, that dew formed in clear intervals will often even evaporate again when the sky becomes thickly overcast (Rule 4. § 150.).

(167.) When we now come to assemble these partial inductions so as to raise from them a general conclusion, we consider, 1st, That all the conclusions we have come to have a reference to that first general fact – the cooling of the exposed surface of the body dewed below the temperature of the air. Those surfaces which part with their heat outwards most readily, and have it supplied from within most slowly, will, of course, become coldest if there be an opportunity for their heat to escape, and not be restored to them from without. Now, a clear sky affords such an opportunity. It is a law well known to those who are conversant with the nature of heat, that heat is constantly escaping from *all bodies* in rays, or by *radiation,* but is as constantly restored to them by the similar radiation of others surrounding them. Clouds and surrounding objects therefore act as opposing causes by replacing the whole or a great part of the heat so radiated away, which can escape effectually, without being replaced, only through openings into infinite space. Thus, at length, we arrive at the general proximate cause of dew, in the cooling of the dewed surface by radiation faster than its heat can be restored to it, by communication with the ground, or by counter-radiation; so as to become colder than the air, and thereby to cause a condensation of its moisture.

(168.) We have purposely selected this theory of dew, first developed by the late Dr. Wells, as one of the most beautiful specimens we can call to mind of inductive experimental enquiry lying within a moderate compass. It is not possible in so brief a space to do it justice; but we earnestly recommend his work[4] (a short and very entertaining one) for perusal to the student of natural philosophy, as a model with which he will do well to become familiar.

(169.) In the analysis above given, the formation of dew is referred to two more general phenomena; the radiation of heat, and the condensation of invisible vapour by cold. The cause of the former is a much higher enquiry, and may be said, indeed, to be totally unknown; that of the latter actually forms a most important branch of physical enquiry. In such a case, when we reason upwards till we reach an ultimate fact, we regard a phenomenon as fully explained; as we consider the branch of a tree to terminate when traced to its insertion in the trunk, or a twig to its junction with the branch; or rather, as a rivulet retains its importance and its name till lost in some larger tributary, or in the main river which delivers it into the ocean. This, however, always supposes that, on a reconsideration of the case, we see clearly how the admission of such a fact, with all its attendant laws, will perfectly account for *every particular* – as well those which, in the different stages of the induction, have led us to a knowledge of it, as those which we had neglected, or considered less minutely than the rest. But, had we no previous knowledge of the radiation of heat, this same induction would have made it known to us, and, duly considered, might have led to the knowledge of many of its laws.

(170.) In the study of nature, we must not, therefore, be scrupulous as to *how* we reach to a knowledge of such general facts: provided only we verify them carefully when once detected, we must be content to seize them wherever they are to be found. And this brings us to consider the *verification* of inductions.

(171.) If, in our induction, every individual case has actually been present to our minds, we are sure that it will find itself duly *represented* in our final conclusion: but this is impossible for such cases as were *unknown* to us, and hardly ever happens even with all the known cases; for such is the tendency of the human mind to speculation, that on the least idea of an analogy between a few phenomena, it leaps forward, as it were, to a cause or law, to the temporary neglect of all the rest; so that, in fact, almost all our principal inductions must be regarded as a series of ascents and descents, and of conclusions from a few cases, verified by trial on many.

(172.) Whenever, therefore, we think we have been led by induction to the knowledge of the proximate cause of a phenomenon or of a law of nature, our next business is to examine deliberately and *seriatim* all the cases we have collected of its occurrence, in order to satisfy ourselves that they are explicable by our cause, or fairly included in the expression of our law: and in case any exception occurs, it must be carefully noted and set aside for re-examination at a more advanced period, when, possibly, the cause of exception may appear, and the exception itself, by allowing for the effect of that cause, be brought over to the side of our induction; but should exceptions prove numerous and various in their features, our faith in the conclusion will be proportionally shaken, and at all events its importance lessened by the destruction of its universality.

(173.) In the conduct of this verification, we are to consider whether the cause or law to which we are conducted be one already known and recognised as a more general one, whose nature is well understood, and of which the phenomenon in question is but one more case in addition to those already known, or whether it be one less general, less known, or altogether new. In the latter case, our verification will suffice, if it merely shows that all the cases considered are plainly cases in point. But in the former, the process of verification is of a much more severe and definite kind. We must trace the action of our cause with distinctness and precision, as modified by all the circumstances of each case; we must estimate its effects, and show that nothing unexplained remains behind; at least, in so far as the presence of unknown modifying causes is not concerned.

(174.) Now, this is precisely the sort of process in which *residual phenomena* (such as spoken of in art. 158.) may be expected to occur. If our induction be really a valid and a comprehensive one, *whatever* remains unexplained in the comparison of its conclusion with particular cases, under all their circumstances, *is* such a phenomenon, and comes in its turn to be a subject of inductive reasoning to discover its cause or laws. It is thus that we may be said to witness facts with the eyes of reason; and it is thus that we are continually attaining a knowledge of new phenomena and new laws which lie beneath the surface of things, and give rise to the creation of fresh branches of science more and more remote from common observation.

(175.) Physical astronomy affords numerous and splendid instances of this. The law, for example, which asserts that the planets are retained in their orbits about the sun, and satellites about their primaries, by an attractive force, decreasing as the square of the distances increases, comes to be verified in each particular case by deducing from it the exact motions which, under the circumstances, ought to take place, and comparing them with fact. This comparison, while it verifies in general the existence of the law of gravitation as supposed, and its adequacy to explain all the principal motions of every body in the system, yet leaves some small deviations in those of the planets, and some very considerable ones in that of the moon and other satellites, still unaccounted for; residual phenomena, which still remain to be traced up to causes. By further examining these, their causes have at length been ascertained, and found to consist in the mutual actions of the planets on each other, and the disturbing influence of the sun on the motions of the satellites.

(176.) But a law of nature has not that degree of generality which fits it for a stepping-stone to greater inductions, unless it be *universal* in its application. We cannot rely on its enabling us to extend our views beyond the circle of instances from which it was obtained, unless we have already had experience of its power to do so; unless it actually *has* enabled us before trial to say what will take place in cases analogous to those originally contemplated; unless, in short, we have studiously placed ourselves in the situation of its antagonists, and even perversely endeavoured to find exceptions to it without success. It is in the precise proportion that a law once obtained endures this extreme severity of trial, that its value and importance are to be estimated; and our next step in the verification of an induction must therefore consist in *extending* its application to cases not originally contemplated; in studiously varying the circumstances under which our causes act, with a view to ascertain whether their effect is general; and in pushing the application of our laws to extreme cases.

(177.) For example, a fair induction from a great number of facts led Galileo to conclude that the accelerating power of gravity is the same on all sorts of bodies, and on great and small masses indifferently; and this he exemplified by letting bodies of very different natures and weights fall at the same instant from a high tower, when it was observed that they struck the ground at the same moment, abating a certain trifling difference, due, as he justly believed it to be, to the greater proportional resistance of the air to light than to heavy bodies. The experiment could not, at that time, be fairly tried with extremely light substances, such as cork, feathers, cotton, &c. because of the great resistance experienced by these in their fall; no means being then known of removing this cause of disturbance. It was not, therefore, till after the invention of the air-pump that this law could be put to the severe test of an extreme case. A guinea and a downy feather were let drop at once from the upper part of a tall exhausted glass, and struck the bottom at the same moment. Let any one make the trial *in the air*, and he will perceive the force of an *extreme case.*

(178.) In the verification of a law whose expression is *quantitative*, not only must its generality be established by the trial of it in as various circumstances as possible, but every such trial must be one of precise measurement. And in such cases the means taken for subjecting it to trial ought to be so devised as to repeat and multiply a great number of times any deviation (if any exist); so that, let it be ever so small, it shall at last become sensible.

(179.) For instance, let the law to be verified be, that *the gravity of every material body is in the direct proportion of its mass*, which is only another mode of expressing Galileo's law above mentioned. The time of falling from any moderate height cannot be measured with precision enough for our purpose: but if it can be repeated a very great multitude of times *without any loss or gain* in the intervals, and the whole amount of the times of fall so repeated measured by a clock; and if at the same time the resistance of the air can be rendered *exactly alike* for all the bodies tried, we have here Galileo's trial in a much more refined state; and it is evident that almost unlimited exactness may be obtained. Now, all this Newton accomplished by the simple and elegant contrivance of enclosing in a hollow pendulum the same weights of a great number of substances the most different that could be found in all respects, as gold, glass, wood, water, wheat, &c.,[5] and ascertaining the time required for the pendulum so charged to make a great number of oscillations; in each of which it is clear the weights had to fall, and be raised again successively, without loss of time, through the same *identical* spaces. Thus any difference, however inconsiderable, that might exist in the time of one such fall and rise would be multiplied and accumulated till they became sensible. And none having been discovered by so delicate a process in any case, the law was considered verified both in respect of generality and exactness. This, however, is nothing to the verifications afforded by astronomical phenomena, where the deviations, if any, accumulate for thousands of years instead of a few hours.

(180.) The surest and best characteristic of a well-founded and extensive induction, however, is when verifications of it spring up, as it were, spontaneously, into notice, from quarters where they might be least expected, or even among instances of that very kind which were at first considered hostile to them. Evidence of this kind is irresistible, and compels assent with a weight which scarcely any other possesses. To give an example: M. Mitscherlich had announced a law to this effect – *that* the chemical elements of which all bodies consist are susceptible of being classified in distinct groups, which he termed *isomorphous* groups;

and *that* these groups are so related, that when similar combinations are formed of individuals belonging to two, three, or more of them, such combinations will crystallize in the same geometrical forms. To this curious and important law there appeared a remarkable exception. According to professor Mitscherlich, the arsenic and phosphoric acids *are* similar combinations coming under the meaning of his law, and their combinations with soda and water, forming the salts known to chemists under the names of arseniate and phosphate of soda, ought, if the law were general, to crystallize in identical shapes. The fact, however, was understood to be otherwise. But lately, Mr. Clarke, a British chemist, having examined the two salts attentively, ascertained the fact that their compositions deviate essentially from that similarity which M. Mitscherlich's law requires; and that, therefore, the exception in question disappears. This was something: but, pursuing the subject further, the same ingenious enquirer happily succeeded in producing a *new* phosphate of soda, differing from that generally known in containing a different proportion of water, and agreeing in composition exactly with the arseniate. The crystals of this new salt, when examined, were found by him to be precisely identical in form with those of the arseniate: thus verifying, in a most striking and totally unexpected manner, the law in question, or, as it is called, the law of isomorphism.

(181.) Unexpected and peculiarly striking confirmations of inductive laws frequently occur in the form of residual phenomena, in the course of investigations of a widely different nature from those which gave rise to the inductions themselves. A very elegant example may be cited in the unexpected confirmation of the law of the developement of heat in elastic fluids by compression, which is afforded by the phenomena of sound. The enquiry into the cause of sound had led to conclusions respecting its mode of propagation, from which its velocity in the air could be precisely calculated. The calculations were performed; but, when compared with fact, though the agreement was quite sufficient to show the general correctness of the cause and mode of propagation assigned, *yet* the *whole* velocity could not be shown to arise from this theory. There was still a *residual* velocity to be accounted for, which placed dynamical philosophers for a long time in a great dilemma. At length La Place struck on the happy idea, that this might arise from the *heat* developed in the act of that condensation which necessarily takes place at every vibration by which sound is conveyed. The matter was subjected to exact calculation, and the result was at once the complete explanation of the residual phenomenon, and a striking confirmation of the general law of the developement of heat by compression, under circumstances beyond artificial imitation.

(182.) In extending our inductions to cases not originally contemplated, there is one step which always strikes the mind with peculiar force, and with such a sensation of novelty and surprise, as often gives it a weight beyond its due philosophic value. It is the transition from the little to the great, and *vice versa,* but especially the former. It is so beautiful to see, for instance, an experiment performed in a watch-glass, or before a blowpipe, succeed, in a great manufactory, on many tons of matter, or, in the bosom of a volcano, upon millions of cubic fathoms of lava, that we almost forget that these great masses are made up of watch-glassfuls, and blowpipe-beads. We see the enormous intervals between the stars and planets of the heavens, which afford room for innumerable processes to be carried on, for light and heat to circulate, and for curious and complicated motions to go forward among them: we look more attentively, and we see sidereal systems, probably not less vast and complicated than our own, crowded apparently into a small space (from the effect of their distance from us), and forming groups resembling bodies of a substantial appearance, having form and outline: yet we recoil with incredulous surprise when we are asked *why* we cannot conceive the atoms of a grain of sand to be as remote from each other (proportionally to their sizes) as the stars of the firmament; and why there may not be going on, in that little microcosm, processes as complicated and wonderful as those of the great world around us. Yet the student who makes any progress in natural philosophy will encounter numberless cases in which this transfer of ideas from the one extreme of magnitude to the other will be called for: he will find, for instance, the phenomena of the propagation of winds referred to the same laws which regulate the propagation of motions through the smallest masses of air; those of lightning assimilated to the mere communication of an electric spark, and those of earthquakes to the tremors of a stretched wire: in short, he must lay his account to finding the distinction of great and little altogether annihilated in nature: and it is well for man that such is the case, and that the same laws, which he can discover and verify in his own circumscribed sphere of power, should prove available to him when he comes to apply them on the greatest scale; since it is thus only that he is enabled to become an exciting cause in operations of any considerable magnitude, and to vindicate his importance in creation.

(183.) But the business of induction does not end here: its final result must be followed out into all its consequences, and applied to all those cases which seem even remotely to bear upon the subject of enquiry. Every new addition to our stock of causes becomes a means of fresh attack with new vantage ground upon all those unexplained parts of former phenomena which have resisted previous efforts. It can hardly be pressed forcibly enough on the attention of the student of nature, that there is scarcely any natural phenomenon which can be fully and completely explained in all its circumstances, without a union of several, perhaps of all, the sciences. The great phenomena of astronomy, indeed, may be considered exceptions; but this is merely because their scale is so vast that one only of the most widely extending forces of nature takes the lead, and all those agents whose sphere of action is limited to narrower bounds, and which determine the production of phenomena nearer at hand, are thrown into the back ground, and become merged and lost in comparative insignificance. But in the more intimate phenomena which surround us it is far otherwise. Into what a complication of different branches of science are we not led by the consideration of such a phenomenon as rain, for instance, or flame, or a thousand others, which are constantly going on before our eyes? Hence, it is hardly possible to arrive at the knowledge of a law of any degree of generality in any branch of science, but it immediately furnishes us with a means of extending our knowledge of innumerable others, the most remote from the point we set out from; so that, when once embarked in any physical research, it is impossible for any one to predict where it may ultimately lead him.

(184.) This remark rather belongs to the inverse or *deductive* process, by which we pursue laws into their remote consequences. But it is very important to observe, that the successful process of scientific enquiry demands continually the alternate use of both the *inductive* and *deductive* method. The path by which we rise to knowledge must be made smooth and beaten in its lower steps, and often ascended and descended, before we can scale our way to any eminence, much less climb to the summit. The achievement is too great for a single effort; stations must be established, and communications kept open with all below. To quit metaphor; there is nothing so instructive, or so likely to lead to the acquisition of general views, as this pursuit of the consequences of a law once arrived at into every subject where it may seem likely to have an influence. The discovery of a new law of nature, a new ultimate fact, or one that even temporarily puts on that appearance, is like the discovery of a new element in chemistry. Thus, selenium was hardly discovered by Berzelius in the vitriol works of Fahlun, when it presently made its appearance in the sublimates of Stromboli, and the rare and curious products of the Hungarian mines. And thus it is with every new law, or general fact. It is hardly announced before its traces are found every where, and every one is astonished at its having so long remained concealed. And hence it happens that unexpected lights are shed at length over parts of science that had been abandoned in despair, and given over to hopeless obscurity.

[. . .]

CHAP. VII.

OF THE HIGHER DEGREES OF INDUCTIVE GENERALIZATION, AND OF THE FORMATION AND VERIFICATION OF THEORIES.

(201.) As particular inductions and laws of the first degree of generality are obtained from the consideration of individual facts, so Theories result from a consideration of these laws, and of the proximate causes brought into view in the previous process, regarded all together as constituting a new set of phenomena, the creatures of reason rather than of sense, and each representing under general language innumerable particular facts. In raising these higher inductions, therefore, more scope is given to the exercise of pure reason than in slowly groping out our first results. The mind is more disencumbered of matter, and moves as it were in its own element. What is now before it, it perceives more intimately, and less through the medium of sense, or at least not in the same manner as when actually at work on the immediate objects of sense. But it must not be therefore supposed that, in the formation of theories, we are abandoned to the unrestrained exercise of imagination, or at liberty to lay down arbitrary principles, or assume the existence of mere fanciful causes. The liberty of speculation which we possess in the domains of theory is not like the wild licence of the slave broke loose from his fetters, but rather like that of the freeman who has learned the lessons of self-restraint in the school of just subordination. The ultimate objects we pursue in the highest theories are the same as those of the lowest inductions; and the means by which we can most securely attain them bear a close analogy to those which we have found successful in such inferior cases.

(202.) The immediate object we propose to ourselves in physical theories is the analysis of phenomena, and the knowledge of the hidden processes of nature in their production, so far as they can be traced by us. An important part of this knowledge consists in a discovery of the actual structure or mechanism of the universe and its parts, through which, and by which, those processes are executed; and of the agents which are concerned in their performance. Now, the mechanism of nature is for the most part either on too large or too small a scale to be immediately cognizable by our senses; and her agents in like manner elude direct observation, and become known to us only by their effects. It is in vain therefore that we desire to become witnesses to the processes carried on with such means, and to be admitted into the secret recesses and laboratories where they are effected. Microscopes have been constructed which magnify more than a thousand times in *linear* dimension, so that the smallest visible grain of sand may be enlarged to the appearance of one a thousand million times more bulky; yet the only impression we receive by viewing it through such a magnifier is, that it reminds us of some vast fragment of a rock, while the intimate structure on which depend its colour, its hardness, and its chemical properties, remains still concealed: we do not seem to have made even an approach to a closer analysis of it by any such scrutiny.

(203.) On the other hand, the mechanism of the great system of which our planet forms a part escapes immediate observation by the immensity of its scale, nay, even by the slowness of its evolutions. The motion of the minute hand of a watch can hardly be perceived without the closest attention, and that of the hour hand not at all. But what are these, in respect of the impression of slowness they produce in our minds, compared with a revolving movement which takes a whole year, or twelve, thirty, or eighty years to complete, as is the case with the planets in their revolutions round the sun. Yet no sooner do we come to reflect on the linear dimensions of these orbs, (which however we do not *see*, nor can we measure them but by a long, circuitous, and difficult process,) than we are lost in astonishment at the swiftness of the very motions which before seemed so slow.[6] The motion of the sails of a windmill offers (on a small scale) an illustrative case. At a distance the rotation seems slow and steady – but when we stand close to one of the sails in its sweep, we are surprised at the swiftness with which it rushes by us.

(204.) Again, the agents employed by nature to act on material structures are invisible, and only to be traced by the effects they produce. Heat dilates matter with an irresistible force; but what heat *is,* remains yet a problem. A current of electricity passing along a wire moves a magnetized needle at a distance; but except from this effect we perceive no difference between the condition of the wire when it conveys and when it does not convey the stream: and we apply the terms current, or stream, to the electricity only because in some of its relations it reminds us of something we have observed in a stream of air or water. In like manner we see that the moon circulates about the earth; and because we believe it to be a solid mass, and have never seen one solid substance revolve round another within our reach to handle and examine unless retained by a force or united by a tie, we conclude that there *is* a force, and a mode of connection, between the moon and the earth; though, what that mode can be, we have no conception, nor can imagine *how* such a force can be exerted at a distance, and with empty space, or at most an invisible fluid, between. (See § 148.)

(205.) Yet are we not to despair, since we see regular and beautiful results brought about in human works by means which nobody would, at first sight, think could have any thing to do with them. A sheet of blank paper is placed upon a frame, and shoved forwards, and after winding its way successively over and under half a dozen rollers, and performing many other strange evolutions, comes out printed on both sides. And, after all, the acting cause in this process is nothing more than a few gallons of water boiled in an iron vessel, at a distance from the scene of operations. But *why* the water so boiled should be capable of producing the active energy which sets the whole apparatus in motion is, and will probably long remain, a secret to us.

(206.) This, however, does not at all prevent our having a very perfect comprehension of the whole subsequent process. We might frequent printing-houses, and form a theory of printing, and having worked our way up to the point where the mechanical action commenced (the boiler of the steam-engine), and verified it by taking to pieces, and putting together again, the train of wheels and the presses, and by sound theoretical examination of all the transfers of motion from one part to another; we should, at length, pronounce our theory good, and declare that we understood printing thoroughly. Nay, we might even go away and apply the principles of mechanism we had learned in this enquiry to other widely different purposes; construct other machines, and put them in motion by the same moving power, and all without arriving at any correct idea as to the ultimate source of the force employed. But, if we were inclined to theorize farther, we might do so; and it is easy to imagine how two theorists might form very different *hypotheses* as to the origin of the power which alternately raised and depressed the piston-rod of the engine. One, for example, might maintain that the boiler (whose contents we will suppose that neither theorist has been permitted to examine) was the den of some powerful unknown animal, and he would not be without plausible analogies in the warmth, the supply of fuel and water, the breathing noises, the smoke, and above all, the mechanical power exerted. He would say (not without a show of reason), that where there is a positive and wonderful effect, and many strong analogies, such as materials consumed, and all the usual signs of life maintained, we are not to deny the existence of animal life because we know no animal that consumes such food. Nay, he might observe with truth, that the fuel actually consists of the chemical ingredients which constitute the chief food of all animals, &c.; while, on the other hand, his brother theorist, who caught a glimpse of the fire, and detected the peculiar sounds of ebullition, might acquire a better notion of the case, and form a theory more in consonance with fact.

(207.) Now, nothing is more common in physics than to find two, or even many, *theories* maintained as to the origin of a natural phenomenon. For instance, in the case of heat itself, one considers it as a really existing material fluid, of such exceeding subtlety as to penetrate all bodies, and even to be capable of combining with them chemically; while another regards it as nothing but a rapid vibratory or rotatory motion in the ultimate particles of the bodies heated; and produces a singularly ingenious train of mechanical reasoning to show, that there is nothing contradictory to sound dynamical principles in such a doctrine. Thus, again, with light: one considers it as consisting in actual particles darted forth from luminous bodies, and acted upon in their progress by forces of extreme intensity residing in the substances on which they strike; another, in the vibratory motion of the particles of luminous bodies, communicated to a peculiar subtle and highly elastic ethereal medium, filling all space, and conveyed through it into our eyes, as sounds are to our ears, by the undulations of the air.

(208.) Now, are we to be deterred from framing hypotheses and constructing theories, because we meet with such dilemmas, and find ourselves frequently beyond our depth? Undoubtedly not. *Est quodam prodire tenus si non datur ultra.* Hypotheses, with respect to theories, are what presumed proximate causes are with respect to particular inductions: they afford us motives for searching into analogies; grounds of citation to bring before us all the cases which seem to bear upon them, for examination. A well imagined hypothesis, if it have been suggested by a fair inductive consideration of general laws, can hardly fail at least of enabling us to generalize a step farther, and group together several such laws under a more universal expression. But this is taking a very limited view of the value and importance of hypotheses: it may happen (and it has happened in the case of the undulatory doctrine of light) that such a weight of analogy and probability may become accumulated on the side of an hypothesis, that we are compelled to admit one of two things; either that it is an actual statement of what really passes in nature, or that the reality, whatever it be, must run so close a parallel with it, as to admit of some mode of expression common to both, at least in so far as the phenomena actually known are concerned. Now, this is a very great step, not only for its own sake, as leading us to a high point in philosophical speculation, but for its applications; because whatever conclusions we deduce from an hypothesis so supported must have at least a strong presumption in their favour; and we may be thus led to the trial of many curious experiments, and to the imagining of many useful and important contrivances, which we should never otherwise have thought of, and which, at all events, *if* verified in practice, are real additions to our stock of knowledge and to the arts of life.

(209.) In framing a theory which shall render a rational account of any natural phenomenon, we have *first* to consider the agents on which it depends, or the causes to which we regard it as ultimately referable. These agents are not to be arbitrarily assumed; they must be such as we have good inductive grounds to believe do exist in nature, and do perform a part in phenomena analogous to those we would render an account of; or such, whose presence in the actual case can be demonstrated by unequivocal signs. They must be *veræ causæ*, in short, which we can not only show to exist and to act, but the laws of whose action we can derive independently, by direct induction, from experiments purposely instituted; or at least make such suppositions respecting them as shall not be contrary to our experience, and which will remain to be verified by the coincidence of the conclusions we shall deduce from them, with facts. For example, in the theory of gravitation we suppose an agent, – *viz.* force, or mechanical power, – to act on *any* material body which is placed in the presence of *any* other, and to urge the two mutually towards each other. This is a *vera causa;* for heavy bodies (that is, all bodies, but some more, some less,) tend to, or endeavour to reach, the earth, and require the exertion of force to counteract this endeavour, or to keep them up. Now, that which opposes and neutralizes force *is* force. And again, a plumb-line, which, when allowed to hang freely, always hangs perpendicularly; is found to hang observably aside from the perpendicular when in the neighbourhood of a considerable mountain; thereby proving that a force is exerted upon it, which draws it towards the mountain. Moreover, since it is a fact that the moon does circulate about the earth, it must be drawn towards the earth by a force; for if there were no force acting upon it, it would go on in a straight line without turning aside to circulate in an orbit, and would, therefore, soon go away and be lost in space. This force, then, which we call the *force* of gravity, is a real cause.

(210.) We have next to consider the laws which regulate the action of these our primary agents; and these we can only arrive at in three ways: 1st, By inductive reasoning; that is, by examining all the cases in which we know them to be exercised, inferring, as well as circumstances will permit, its amount or intensity in each particular case, and then piecing together, as it were, these *disjecta membra,* generalizing from them, and so arriving at the laws desired; 2dly, By forming at once a bold hypothesis, particularizing the law, and trying the truth of it by following out its consequences and comparing them with facts; or, 3dly, By a process partaking of both these, and combining the advantages of both without their defects, viz. by assuming indeed the laws we would discover, but so generally expressed, that they shall include an unlimited variety of particular laws; – following out the consequences of this assumption, by the application of such general principles as the case admits; – comparing them in succession with all the particular cases within our knowledge; and, lastly, *on this comparison*, so modifying and restricting the general enunciation of our laws as to *make the results agree.*

(211.) All these three processes for the discovery of those general elementary laws on which the higher theories are grounded are applicable with different advantage in different circumstances. We might exemplify their successive application to the case of gravitation: but as this would rather lead into a disquisition too particular for the objects of this discourse, and carry us too much into the domain of technical mathematics, we shall content ourselves with remarking, that the method last mentioned is that which mathematicians (especially such as have a considerable command of those general modes of representing and reasoning on quantity, which constitute the higher analysis,) find the most universally applicable, and the most efficacious; and that it is applicable with especial advantage in cases where subordinate inductions of the kind described in the last section have already led to laws of a certain generality admitting of mathematical expression. Such a case, for instance, is the elliptic motion of a planet, which is a general proposition including the statement of an infinite number of particular *places*, in which the laws of its motion allow it to be some time or other found, and for which, of course, the law of force must be so assumed as to account.

Notes

1 Lyell's Principles of Geology, vol. i. Fourrier, Mém. de l'Acad. des Sciences, tom. vii. p. 592. "L'établissement et le progrès des sociétés humaines, l'action des forces naturelles, peuvent changer notablement, et dans de vastes contrées, l'état de la surface du sol, la distribution des eaux, et les grands mouvemens de l'air. De tels effets sont propres à faire varier, dans le cours de plusieurs siècles, le dégré de la chaleur moyenne; car les expressions analytiques comprennent des coefficiens qui se rapportent à l'état superficiel, et qui influent beaucoup sur la valeur de la température." In this enumeration, by M. Fourrier, of causes which may vary the general relation of the surface of extensive continents to heat, it is but justice to Mr. Lyell to observe, that the gradual shifting of the *places* of the continents themselves on the surface of the globe, by the abrading action of the sea on the one hand, and the elevating agency of subterranean forces on the other, does not expressly occur and cannot be fairly included in the general sense of the passage, which confines itself to the consideration of such changes as may take place on the existing surface of the land.

2 The reader will find this subject further developed in a paper lately communicated to the Geological Society.

3 Phil. Trans. 1824.

4 Wells on Dew.

5 Principia, book iii. prop. 6

6 "When we are told that Saturn moves in his orbit more than 22,000 miles an hour, we fancy the motion to be swift; but when we find that he is more than three hours moving his own diameter, we must then think it, as it really is, slow." Thirty Letters on various Subjects, by William Jackson, 1795.

8

[REVIEW OF] *A PRELIMINARY DISCOURSE ON THE STUDY OF NATURAL PHILOSOPHY*

William Whewell

Source: William Whewell, "[Review of] *A Preliminary Discourse on the Study of Natural Philosophy*," *The Quarterly Review*, 45:90 (1831), pp. 374–391, 398–402.

Art. III. – *A Preliminary Discourse on the Study of Natural Philosophy.* By J. F. W. Herschel, Esq., M.A. of St. John's College, Cambridge. (*Published in Dr. Lardner's Cabinet Cyclopædia.*) London. 1830.

THE history and philosophy of physical science are attractive topics to many to whom science itself appears thorny and repulsive; and to present these in a manner at the same time lucid and profound, is justly considered as a task demanding a tone and expansion of mind which the successful cultivators of such knowledge in its details do not always possess. We need not affect to undervalue our own vocation; but, at the same time, we are as ready as others to allow that it is eminently fortunate for the world when men of high scientific eminence, and of large speculative views, are led to promulgate, in a distinct and separate form, their notions of the general character and bearing of those portions of human knowledge into which they have the clearest insight. Such surveys may naturally take a far wider range, and offer a more connected picture, than critiques of particular books or authors can do; and the name of a writer who is principally known by abstruse and technical writings, unapproachable by the common reader, gives a peculiar charm to his less professional views of the philosophy and history of his studies; for these, by bringing before us principles and opinions in which he appeals to our common sympathy and understanding, seem to admit us within the barrier with which the passionless discussions and repulsive formularies of science generally surround him.

The present generation may be held to have been fortunate in obtaining from some of the greatest of the great men who have lived in it, expositions of this character. It would be difficult to find two names higher in European reputation than those of Cuvier and Berzelius. From the former we had, in 1808, his 'Histoire des Progrès des Sciences Naturelles depuis 1789 jusqu'à ce jour;' and this work may be considered as

DOI: 10.4324/9781003263814-10

continued to our own times by the annual reports on the same subject, and *éloges* of deceased academicians, which his office of Secretary to the Institute has elicited from him; and in the lectures which he is, we believe, at present delivering in Paris, he is supplying, in a manner in which no other person could do it, the still earlier history of these provinces of knowledge. Berzelius has, ever since 1822, published an 'Annual Report on the Progress of the Physical Sciences;' and has in this, with a knowledge and candour almost peculiar to him, endeavoured to do justice to the yearly additions made to the stores of the scientific community. England has not hitherto contributed her share to this kind of literature from pens of so much authority; but it is unnecessary to dwell on a deficiency which (thanks to the *Cabinet Cyclopædia,* and the *Family Library*) there is every reason to think will now be speedily supplied.[1]

So far as the remarkable volume before us belongs to the kind of review of which we have spoken (which, however, is only one of its objects), it is worthy to stand by the side of the productions of Cuvier and Berzelius; exhibiting an intimate acquaintance with modern research, and a most scrupulous and active spirit of justice, like the qualities which characterize the Swedish chemist; along with a striking mode of presenting recondite truths, and an imaginative vivacity, which has its charm no less than the brilliant eloquence of the great naturalist of France. Mr. Herschel is, indeed, a person eminently qualified for this task, being a distinguished example of a person possessing a profound and complete knowledge of almost every branch of physics; and thus an instance of a combination which, amidst the modern multiplication of the divisions and details of these studies, is becoming rare. Having been thoroughly disciplined in all that can form the character of an accomplished man of science, he has shown himself with singular alacrity at every point of the frontier of human knowledge, where there seemed to be a chance that the boundary line might be pushed forwards. In pure mathematics he very early established his name as one of the first mathematicians in England; and might have been one of the first in Europe, if he had not been more strongly drawn to other pursuits. In the singularly splendid and striking researches of physical optics, one of the pages of discovery as yet but half unfolded, he has placed himself among the *very* small number of those who have both added important experimental laws to those which were previously known, and have weighed the relation of these discoveries to the refined and recondite theory towards which they seem to point. In chemistry, his researches on the hyposulphites, – in galvanism, his examination of the motions of fluid conductors, – exhibit the same readiness to press forward with the advance; and we need hardly remind our readers that by far the most striking fact which has been for centuries added to our knowledge of the heavens – the rotation of the double stars, – first caught sight of by one Herschel, has owed its final and distinct confirmation to another. We might add acoustics and magnetism, mineralogy and geology, to the subjects on which Mr. Herschel's mind and hands have been energetically employed. He has, therefore, no slight or ordinary title to be listened to, when he delivers his opinions on the present condition and mutual relations of the provinces of our knowledge of external nature.

In point of fact, however, the 'Introduction to the Study of Natural Philosophy' is much more than a mere survey of the state of such knowledge. The book contains another important element, viz. Mr. Herschel's view of the philosophy of physical science, the principles on which its structure rests, the maxims by which its researches have been and must be successfully conducted. Opinions on such a subject from a person who has not only *verified* his principles over a wide range of speculation, but *used* them with practical success, are likely to be read with interest. There is another reason which makes this part of the work an object of curiosity. It is, we believe, one of the first considerable attempts to expound in any detail the rules and doctrines of that method of research to which modern science has owed its long-continued steady advance and present flourishing condition. It is certainly strange that this should be so; that while volumes upon volumes have been written upon the nature of human knowledge and the laws of human thought, this class of speculations should never have been constructed with a peculiar reference to those mental processes which have been exemplified in the progress of modern science – the most undisputed and the most admired assemblage of truths which has ever yet been obtained. We hope that the day will come when this shall be otherwise: when the received 'philosophy of mind' shall have its leading chapters devoted to the most successful employments of the intellect: when the received 'art of reasoning' shall contain something more than rules for deriving the consequences of assumed principles; and when 'definitions' shall no longer be considered as the fountains of what we can know with regard to things altogether independent of the operations of human thought.

It is the more remarkable that our prevalent metaphysics and logic should be so inapplicable to physics, when we observe that the whole atmosphere of literature rings with the name of '*Lord* Bacon,' and with the pæans of 'Inductive Philosophy.' To many of our readers, discussions on the practical meaning and exemplification of this philosophy may be attractive; but we fear there is a large class of respectable and intelligent persons who are absolutely wearied with the phrase, and the unsubstantial declamation which has, as they conceive, been connected with it. Yet perhaps even these readers may allow it to be a benefit if we can, with Mr. Herschel's assistance, fix this, and similar terms, to a settled meaning and definite limits. Some time back, the eminent person who now, as a scientific Lord Chancellor, is peculiarly Bacon's successor, expressed a hope that in the course of the schoolmaster's triumphs, the day would come when every Englishman would be able to read Bacon; Mr. Cobbett, who has long maintained that the true interest of Englishmen is to keep pigs and read his Register, observed somewhat contemptuously, that it would be more to the purpose to hope we might all come to *eat* bacon. We shall not attempt to decide what ought in this case to be the proportion between those who read and those who eat, but we think it clearly desirable that both parties should digest their food; and leaving the masticators, with our best wishes, to the exercise of their gastric energies, we will offer a few remarks for the benefit of the other class.

Mr. Herschel's work is indeed an admirable comment upon the 'Novum Organum Scientiarum;' and it possesses this character, not only when he is professedly adopting and applying Bacon's views, but often when he is pursuing trains of reflection purely suggested by his own knowledge of the history and relations of science. The two works derive mutual illustration from being considered together; and such a circumstance cannot but give us an elevated impression both of Bacon's far-seeing sagacity, and of his successor's comprehensive and philosophical spirit.

Mr. Herschel's discourse is divided into three parts, of which the first may be described as referring to the dignity and uses of physical science; the second to its general principles and method; the third to the peculiar principles of the different departments of science, their causes and bearings. We shall probably make our observations more perspicuous by explaining, in the first place, the peculiar nature of modern science, and of that Inductive or Baconian method to which its marvels are attributed.

If we are to judge from the various explanations given of the term *induction,* whenever it is attempted to be defined, the prevalent conceptions of its signification are sufficiently vague and inconsistent. We by no means pretend to object to any one's using the term in a sense different from ourselves, and least of all if he can so express any valuable truth.[2] But we are here to employ it to designate that process which is the essential element of the physical sciences as they now exist, and in virtue of which they are called *inductive* sciences. For such a purpose, induction *must* be described to be 'the process of considering a class of phenomena, or two associated classes of phenomena, as represented by a general *law,* or single conception of the mind;' such a single law, or representation, is capable of including so many phenomena, in virtue of the natural connexions of our thoughts; and will derive its validity from its being a connected representation of the phenomena, and nothing more. In this manner Induction agrees with mere Observation in accumulating facts, and with pure Reason in stating general propositions; but she does *more* than Observation, inasmuch as she not only collects facts, but catches some connexion or relation among them; and *less* than pure Reason (at least in the same way), because she only declares that there *are* connecting properties, without asserting that they *must* exist of necessity and in all cases. If we conceive the facts of external nature to lie before us like a heap of pearls of various forms and sizes, mere Observation takes up an indiscriminate handful of them; Induction seizes some thread on which a portion of the heap are strung, and binds such threads together; while Deduction, the purely reasoning faculty, employs herself in measuring spaces on the naked thread, and in devising patterns which can only be of use so far as the jewels will fit the places thus assigned them. Thus, to take a very ancient instance of an induction, in which phenomena were connected by means of the conception of their succession in *time:* it was very early discovered that the eclipses of the sun and moon recurred in the same order and at the same intervals as at first, after a period of a little more than eighteen years, viz. after 223 lunations. This ancient cycle, the Chaldean *saros,* thus gave a connected and distinct representation of this series of phenomena, which, without it, appear so lawless and irregular; and that this succession really obtains, is an inductive truth, which, in the course of time, led to other important steps in lunar astronomy. Again, the rate at which the air is rarefied by heat is found, from experiment, to be properly *represented* by saying that the elasticity, or resistance to compression, is *proportional* to the thermometrical increase of temperature. Here is an inductive law of the connexion of phenomena. Such truths, so established, may again be *included* in others of a more general character and higher order; and in this manner we may have a vast structure of science built up, which consists indeed only of phenomena as its materials, but in which phenomena are cemented by these successive modes of connected representation, and form a series of laws each supporting other laws more general; so that we obtain, combined with all the physical reality of the external world, all the permanence and symmetry of mathematical relations.

That this exhibition of mere phenomena, and of pure ideal conceptions, as in harmony with each other, does constitute the supreme success of the progress of scientific inquiry, we believe will be most clear to the students of those sciences which have been carried to the farthest limit of generality and simplification; as, for instance, mechanics and physical astronomy. But the same description will be found to apply, though less obviously perhaps, to other sciences, which have as yet been obliged to stop short in this advance.

Mr. Herschel's account of the matter agrees with that which we have given. After speaking of the importance of classifying facts and objects, he says –

> 'When we have amassed a great store of such *general facts,* they become the objects of another and higher species of classification, and are themselves included in laws which, as they dispose of groups, not individuals, hare a far superior degree of generality, till at length, by continuing the process, we arrive at *axioms* of the highest degree of generality of which science is capable. This process is what we mean by induction.'

When we have attained by this process to propositions of any degree of generality in science, we can often, by pure reasoning, deduce from them, consequences recommended by their curiosity or their utility. Thus, when we have obtained, by induction from observation of which we have given an instance, the laws of the moon's motion in the heavens, we may from these discover at what times, and with what circumstances, eclipses of the sun and moon will occur. And the more general are the truths which we have reached by our inductive ascent, the more copious and varied will be the inferences which we may obtain by reasoning downwards from them. This mutual dependence and contrast of induction and deduction, this successive reasoning up to principles and down to consequences, is one of the most important and characteristic properties of true science. The great legislator of this science ever and anon calls our attention to this feature of his method: 'Neque enim in plano via sita est, sed ascendendo et descendendo; ascendendo ad axiomata, descendendo ad opera.'

It was necessary to premise thus much concerning the nature and principles of the inductive philosophy, in order that the observations which we have to make, or to borrow from Mr. Herschel, concerning its history and rules, might be rendered more generally intelligible.

A circumstance which is sufficiently curious in the history of this philosophy, is this: – That the extraordinary man who promulgated its precepts, at a time when scarcely a single exemplification of them existed, with a precision and sagacity which centuries of experience have hardly enabled us to improve upon, is generally understood to have failed altogether in his attempts to apply his own method to particular cases. There are, no doubt, persons who have a pleasure in believing that this is so, and who examine no further; for there are persons who have a pleasure in finding contrasts and deficiencies in anything which has excited the admiration of the world. Such men are never weary of repeating Pope's 'greatest, wisest, meanest of mankind,' as if to show with how much foresight Bacon commended himself 'to men's charitable speeches:' they would probably have sympathized with another of our poets in his delight at finding *historical* inaccuracies in some of Bacon's political *maxims,* and would have complacently exclaimed with him, 'Such are the sages!' But without trespassing upon such contemplations, we conceive it may be interesting and instructive to point out what really was the nature and cause of Bacon's failure.

That Bacon was a most sagacious and intelligent deviser of experiments, has not been hitherto sufficiently noticed. Mr. Herschel has quoted a number of remarkable instances in which he has suggested modes and subjects of observation eminently adapted to lead to discovery. Many of these have long been put in practice; some have recently been applied for the first time. We might easily add other instances of the same kind: and it is indeed impossible to read the 'Novum Organum,' and the various appendages to it, without perceiving that the author's mind was as unwearied in collecting, as it was profound in comparing. Bacon, then, did not fail after the manner of a theorist who speculates much, having observed little. But he attempted to use his materials; in particular, he instituted an inquiry into the nature of *heat,* which he offers as an exemplification of the method of induction. In this attempt, after making an extensive collection of facts, and performing upon them the various processes which he had enjoined as steps in the method of sound investigation, he brings out at last a result, which it would be foolish in his admirers to pretend to invest with any substantial or permanent value, that 'Heat is an expansive motion, restrained, and exerting itself through the smaller particles of a body, the expansion being towards every side, but with somewhat of an inclination to the upper parts.'

No one can read with attention this part of the 'Novum Organum,' without perceiving that Bacon felt the necessity of giving *some* example to illustrate the method he was recommending, and constructed the *Inquisitio in formam calidi,* for this purpose. Now it is not allowed, even to a mind like Bacon's, to discover general laws of nature merely because he wishes to do so; or rather such a mind, formed to prescribe the rules by which discovery was to be governed, was widely distinct from the intellectual cast which executes the detail of this labour. The two characters differ like that of a great general and of a good engineer. This will go far to explain his want of success. But, moreover, we are compelled to say, that the *Inquisitio* is not a judicious application of the inductive method, and that the Master, in illustrating the forms of process which he had devised, has not well exemplified the spirit by which his method is dictated. We hope our readers will think this circumstance worthy a brief illustration.

Bacon's error in the 'Inquisitio' consists in this: that instead of beginning from classes of phenomena, which a fair contemplation of nature shows to be of the same kind, he takes an arbitrary assemblage of phenomena, or at least one which, if not arbitrary, is brought together by a number of loose analogies and accidental associations; namely, all the cases to which the term *hot* is in common language applied. Thus, among the instances which he collects, as materials for his first vindemiation (to use his own term), are the heat of the sun, of common fire, of lightning, of percussion, of fermentation; animal heat, its increase from exercise and from fever; woollen cloths, hot spices, caustic acids, quick lime. In attempting to find, among a vast multitude of such facts, something which is common to all the cases, and peculiarly distinctive of the quality of heat in which they are supposed to agree, it is no wonder that he arrives at nothing but a vague and useless formula. This unfruitful mode of beginning physical researches from a word, was, however, by no means peculiar to Bacon or to his style of speculation. Practical experimentalists, about the same time, took a road not much different. Thus Boyle has his researches into the origin of fluidity and firmness; of volatility and fixedness. Perhaps this was an effect of the natural influence of words; perhaps it was attributable in part to the recent sway of the Aristotelian philosophy, which, like an expelled conqueror, left the traces of its empire in the language of the country, even when its direct authority was no longer acknowledged.

But whatever was the cause of these infant wanderings of experimental philosophy, it will, we think, be worth while to observe what the case was when the course of true speculation began to run smooth. What have been the *inquisitiones de naturâ calidi,* up to our own time, which have led to valuable and abiding truths? The answer to this inquiry will necessarily occupy some space, but we are persuaded that those who feel any interest in such speculations will not think the narrative misplaced. The instance is a most remarkable part of Bacon's works, and has never, so far as we know, been examined by the light of our more advanced knowledge: and the truths really gathered in this field are very fine examples of a more matured 'vindemiation' than it could yield in his day. It will appear too, that though this has been a prosperous path of research, we are, even yet, far from having arrived at any common *form* or *law* of *hot,* such as Bacon wished to reach by a few steps from the phenomena. And in the course of this examination our admiration for our teacher must increase rather than diminish: for his sagacious insight into the form and character of the then remote sciences can hardly be made less remarkable by any inaccuracy of view concerning the length of the path by which they were to be attained.

The inquiries concerning the nature of heat have led to various lines of discovery, which, with Mr. Herschel's assistance, we shall point out.

Instead of taking the vague and unmanageable, because unmeasurable, phenomena of the *feeling* of heat, it was soon perceived that it was necessary to assume, for the standard with which other facts might be compared, some general effect of heat capable of numerical estimation. The expansion which heat produces in certain fluids, as quicksilver, air, spirit of wine, was fixed upon. The thermometer was constructed by Cornelius Drebell; its scale had fixity conferred upon it by Newton; and the laws of temperature, thus measured, speedily exhibited very curious and curiously connected specimens of genuine induction.

The expansion of aëriform fluids by heat, one of the most prominent phenomena, was found to be governed by a simple law, and appeared at first to be in the proportion to the mercurial temperature. It has since been shown, by Mr. Dalton and Dr. Ure, that, in the case of steam, this law requires some correction, and that the increase of elastic power goes on much more rapidly than the increase of temperature; and even for air, there is, at elevated temperatures, a considerable deviation from strict proportionality.

One of the laws recently discovered belonging to this branch of science, is that obtained about the same time by Mr. Dalton and M. Gay-Lussac, independently of each other; viz. the fact that an *equal dilatability* subsists in all gases and vapours, 'one of the most remarkable points in their history,' as Mr. Herschel observes.

The celebrated laws of *latent heat* form another of the lines of successful investigation belonging to this subject. Dr. Black had the merit of ascertaining, that when bodies pass from a solid to a fluid, or from a fluid to a gaseous state, the heat which is applied to them no longer increases their thermometrical temperature, as it would do were no such change to occur, but becomes latent, and appears to be absorbed in maintaining the body in its new state.

We have not room to trace the progress of a nearly related branch of inquiry, that concerning the *specific heat* of substances; or the quantity of heat required to raise their temperature one degree. This has been pursued by Irvine, Lavoisier and Laplace, Crawfurd, Delaroche and Berard, Dulong and Petit, and has led to a very remarkable law – 'that the atoms of all the simple chemical elements have the same capacity for heat.' – *Herschel,* p. 323.

Another set of facts regarding heat have been reduced to the *laws of radiation,* according to which all bodies throw out their heat in all directions, so that hot ones become cold, and cold ones hot; and this in a manner regulated variously according to the material, surface, and other circumstances of the body. These laws have been subjected to a beautiful experimental investigation by MM. Dulong and Petit, and invested in two different theories by Mr. Prevost and Professor Leslie.

Another set of facts, generally mixing with and modifying the action of the last set, but capable of being disentangled and distinctly investigated, are those referable to the *laws of conduction,* by which a body heated in one part conducts or diffuses its heat through the whole of its substance. These laws, as well as the former, have been traced into those mathematical forms and relations by which all things depending on space and number must be regulated; and thus expressed, they require the most refined and difficult analysis to deduce from them those consequences which we are to compare with experiment. This task has, as Mr. Herschel observes (p. 316), 'been executed with admirable success by the late Baron Fourrier, whose recent lamented death has deprived science of an ornament it could ill spare, thinned as its ranks have been within the last few years.'

Another curious set of laws are those which regulate the *transmission* of radiant heat through *transparent* bodies. The heat of our fires is intercepted and detained by screens of glass, and, being so detained, warms them; while solar heat passes freely through and produces no such effect.

> 'The more recent researches of Delaroche, however, have shown that this detention is complete only when the temperature of the source of heat is low; but that, as that temperature is higher, a portion of the heat radiated acquires a power of penetrating glass; and that the quantity which does, so bears continually a larger and larger proportion to the whole, as the heat of the radiant body is more intense. This discovery is very important, as it establishes a community of nature between solar and terrestrial heat; while at the same time it leads us to regard the actual temperature of the sun as far exceeding that of any earthly flame.' – p. 315.

The laws of the *refraction* of the rays of heat, noticed by Sir William Herschel, and of their *polarization,* established by Berard, are also remarkable portions of our knowledge on this subject.

All these facts and laws belong to heat as distinguished from fire or the action of combustion; with regard to these latter striking phenomena, the attempts to establish general laws, exactly and purely representing the phenomena, have, perhaps, been less successful. Yet in this portion of nature, many curious and, no doubt, well-founded connexions have been pointed out; and ever since chemistry has aspired to a philosophical character, the theory of combustion has been laboured more than any other point: it has always been a sort of central fire about which chemical doctrines have revolved. The Stahlian theory of phlogiston, though found wanting when weighed in the *balance,* was a happy seizure of a common principle in facts, really, but not obviously, analogous, – those of combustion, calcination, and acidification; though it erred in attributing these processes to the dissipation of a peculiar ingredient. The new theory, victoriously brought forward by Lavoisier, preserved and extended these analogies, and represented them more conformably to their true nature, as the *addition* of the pervading element, oxygen. Yet this theory was not competent to provide any general rule by which we might define the cases in which light and heat should accompany this union. Berzelius and others have attempted to establish an analogy between the discharge of an electrified body and the light and heat thus evolved in chemical combination; but though this analogy is ingeniously supported, we may at least assert, that it is not yet so clearly defined and traced into the circumstances of each example, that we can justly deem it a law inductively established.

But although, with regard to vivid combustion, we are as yet unable to offer any proper inductive representation of the facts, we are not without some rules to which we can reduce even so subtle and violent an agent as flame. One, which its application has made remarkable, is that discovered by Davy, that ignited vapour cannot pass through minute apertures, preserving its ignition. This result, indeed, was referred by the discoverer to the very elevated temperature of the burning gas, and the consequent cooling which it experiences in its contact with the screen; and the law of minute apertures thus became a consequence of the laws of conduction. But a further consequence, a practical application of the truth thus derived, has supplied Mr. Herschel with one of his strongest examples of the manner in which Nature, who has surrounded man with dangers and inconveniences, ministers protection and remedies through the hands of science, – we allude to 'the safety-lamp, which enables us to walk with light and security, while surrounded with an atmosphere more explosive than gunpowder.' We have an instance in this invention, such as abound in the history of science, that without being able to ascend to axioms, to reach the highest point of the inductive pyramid, we may yet climb far enough to descend again to works of vast utility and unhoped-for power.

Without going further in the enumeration of the researches concerning heat which have been prosecuted since the time of Bacon, it may be collected from the sketch of discovery on this subject which has now been given, how extensive an advance true science has made, each step having been an inductive process, and yet how far we still are from that supreme law or form of heat, which the teacher thought was almost within his reach. It is ever thus in the progress of inductive philosophy, as in the wanderings of the Alpine traveller. The hill which at starting seemed the only barrier between him and his journey's end, serves but as a point from which he descries higher mountains to climb, and wider fields to traverse; but still every step is an ascent, every new prospect is a gain; and the august forms which surround him, and the pure atmosphere into which he rises, make his toil a sport and his perseverance a delight.

Though, as we have said, we cannot as yet combine all the above-described partial laws concerning heat into any one more general law, we are, however, able perhaps to include some of them in a wider representation, which, so far as it is well founded, is a generalization of the next order, and which we will therefore mention.

The laws concerning the radiation, conduction, transmission, refraction, and polarization of heat, are capable of being exactly represented by conceiving it to consist of a peculiar elastic *fluid,* the parts of which act upon each other, and upon the particles of other bodies, by forces governed by certain rules, – these rules being so determined as to make the results calculated from them agree with the phenomena which are observed. We adequately include, therefore, the laws just enumerated, in the assertion that heat is such a fluid; and to the fluid thus conceived the name *caloric* is given. It is not so obviously easy to express the facts of expansion by heat, in agreement with such a conception: to include those of latent heat is perhaps more difficult still. There is, however, no manifest impossibility in reconciling the course of these phenomena with the existence of such a fluid.

Indeed there are several facts, as, for instance, the production of heat by sudden compression, of which this hypothesis is perhaps the most natural account; and these considerations have so far prevailed, that the existence of this fluid caloric, or, in other words, the materiality of heat, has been generally assented to by most modern philosophers. Our readers will perceive, therefore, that the more mature progress of induction has led to a conclusion the reverse of that obtained by the solitary effort of the founder of the method. Bacon conceived heat to be *motion;* the moderns maintain it to be *matter.*

But we should much deceive ourselves if we were to imagine that the march of induction here is terminated or completed. A large body of phenomena are still insubordinate to its authority. What properties must we attribute to our fluid caloric, which may duly bring forth the circumstances of ignition, combustion, explosion, the heat of chemical combination, of electricity, and galvanism? All this is yet to be done; and at this point our inductive process at present stops.

If we have been at all successful in setting forth the history of the modern 'Inquisitiones de naturâ calidi,' the reader will see Bacon's attempt in its true light, and will also have obtained some view of the successive steps of generalization by which the inductive philosopher hopes to ascend from the most multiplied and complex phenomena, through various stages of connecting laws, to a few supreme and simple axioms: such, for instance, as the laws of motion, or the law of universal gravitation, which may properly be said to include a *world* of particular facts.

But there is also another process belonging to this philosophy which has made its appearance in our narrative, and on which we must say a word, – that of *deduction.* From such laws as we have described, the circumstances being given, the facts of each particular case may be deduced; and often, the end being proposed, the fittest means may be assigned; and this process belongs to those reasoning faculties of the mind which lead us from definitions and principles to their consequences. This is an important part of every science when much advanced. But it must carefully be recollected that this process of deduction is only so far applicable as we suppose that of induction to be already successfully performed. The two movements are exactly the reverse of each other: one reasons *to* principles, the other *from* them. The one, beginning its progress from the facts of the external world, travels towards propositions successively more and more comprehensive, and commonly more and more simple: the other commences its course from simple and general assumptions, and derives conclusions limited and complicated by the conditions of particular cases. The difference might seem to be sufficiently obvious; yet it has been so often overlooked, or misapprehended, that we feel greatly obliged to Mr. Herschel for having stated it so distinctly as he has done; and we would willingly hope that, for the future, the distinction will never be lost sight of; that sciences, whether physical, political, or moral, of which the whole structure consists in the consequences of a few axioms, and the whole process in syllogistic logic, or what is equivalent to it, will not hereafter claim the name of inductive sciences, so as to confound the *organ* of Aristotle with that of his reformer.

There is one feature in the deductive part of science which cannot but draw our notice – its mathematical character. The supreme laws, the axiomata generalissima at which we arrive, are *quantitative* laws, and consequently are identical with formulae dependent on space and number. The consequences of these laws applied, repeated, and combined, will therefore involve mathematical relations, and, it may be, in a very complex form. In these cases, the results of our laws are so far from being attainable by the ordinary working of the intellect, that frequently the whole strength of abstract science must be tasked to form the chain of deduction. We have mentioned one instance of this occurrence in speaking of the laws of conduction of heat: we find others in the investigations of physical optics; of the distribution of electricity at the surface of conductors; of the motion of fluids. In these researches it has been absolutely necessary to enlarge the boundaries of mathematics, to invent new modes and branches of calculation, in order to meet the requisitions of physics. In some cases, the relations of space and number have led to dependences so complex, that they have resisted all efforts to develop their consequences; and the career of generalization has been brought to a stand by the unmanageable form of its abstract vehicle. The most illustrious instance of the occurrence of such obstacles, and of their being surmounted, is to be found in Physical Astronomy, where the general law of gravitation being irremoveably established by Newton, the vast task of deducing all its effects was spread out before mathematicians, and has been the subject of their unremitting labour from that time to the present hour. Mr. Herschel has given a lively sketch (p. 273) of the progress of this investigation. Newton had attempted to impress on these reasonings the character of the ancient mathematics; and, though the first inventor of the more powerful analytical method, he had endeavoured to keep this new agent in a kind of bondage to geometry. The continental mathematicians, opposed by no such restraints, pushed forwards in the career of mathematico-physical discovery with emulous rapidity; while in England it was soon found that 'the geometry of Newton was like the bow of Ulysses, which none but its master could bend, Mr. Herschel will describe better than we can the remaining course of these labours.

> 'The legacy of research which Newton may be said to have left to his successors was truly immense. To pursue, through all its intricacies, the consequences of the law of gravitation; to account for all the inequalities of the planetary movements, and the infinitely more complicated, and to us more important ones, of the moon; and to give, what Newton himself certainly never entertained a conception of, a demonstration of the stability and permanence of the system, under all the accumulating influence of its internal perturbations; this labour and this triumph were reserved for the succeeding age, and have been shared in succession by Clairaut, D'Alembert, Euler, Lagrange, and Laplace. Yet, so extensive is the subject, and so difficult and intricate the purely mathematical inquiries to which it leads, that another century may yet be required to go through with the task. The recent discoveries of astronomers

> have supplied matter for investigation, to the geometers of this and the next generation, of a difficulty far surpassing anything that had before occurred. Five primary planets have been added to our system; four of them since the commencement of the present century; and these, singularly deviating from the general analogy of the others, and offering *cases of difficulty* in theory which no one had before contemplated. Yet even the intricate questions to which these bodies have given rise seem likely to be surpassed by those which have come into view with the discovery of several comets revolving in elliptic orbits, like the planets, round the sun, in very moderate periods. But the resources of modern geometry seem, so far from being exhausted, to increase with the difficulties they have to encounter, and already, among the successors of Lagrange and Laplace, the present generation has to enumerate a powerful array of names, which promise to render it not less celebrated in the annals of physico-mathematical research than that which has just passed away.' – pp. 274, 275.

We trust we have given a distinct picture of the combination of induction and deduction, the complete cycle of ascent and descent, which forms the entire scheme of a perfect science, from its first origin to its final development. We are, however, very far from being able to assert that all the departments of physics cultivated by the moderns have completed this cycle, and reached their consummation; on the contrary, the sciences are in very varied points of their progress. They have long been understood to form a sisterhood, and the different individuals of the group offer to us all gradations of growth, from the tottering girl to the full-formed matron. Some of them, indeed, have been long established in the world, and can look for no further advancement, except an increase of their progeny; others appear to experience all the restlessness of youthful expectation and hope of change; others, again, have not yet left the nursery and the spelling-book. To arrange the sciences according to this order of growth, will probably interest some of our readers; but before we proceed to do this, there is one step in their formation of which we must say a word; the more so as it is one very necessary and important process in the inductive ascent, and has yet been entirely omitted by Bacon in his survey. Mr. Herschel has duly weighed its essential necessity, as, with the history of the sciences before him, he could not but do, and has given rules for its conduct.

The step of which we speak is the marking and fixing our generalizations by imposing exact *terms* – a process which may be included in the word *terminology.* As soon as a law of nature has been obtained, it is desirable to express it in some precise and brief phraseology. This once done, a word or a phrase enables us to adduce the law, and thus to render available, at a moment's warning, all the toil and ingenuity which have been embodied in its discovery. Thus the terms we have used, of *conduction, radiation, latent heat, caloric,* are words impregnated, as it were, with a meaning derived from the labours of many observers and calculators. It is by such means that the discoveries and labours of one generation become useful to the next; or rather become inevitably subservient to future discoveries and advances, by being incorporated into the language of each science, and almost imbibed with the air we breathe. Terms of art, well chosen, contain a concentrated essence of all past labours: they connect us, by the shortest road, with the surest elements of knowledge – actual phenomena; by their help we advance through the labyrinth of nature, guided by a thread almost invisible, of which the extremity is fixed to remote but well-known objects; and we go on, knotting it in our turn, to new points, as the windings of our path require. When we speak of the 'magnesian limestone,' or of the 'green sands,' we, by such words, recall identifications established by wide and laborious examinations of fossils, of continuity, of superposition, for which we stand indebted to many able geologists. In chemistry, in like manner, how much is implied in the systematic nomenclature now in common use; and what an important step will be indicated by the change of terminology which seems to be inevitably approaching!

It is remarkable that this fixation of our generalizations by the imposition of names should not have caught Bacon's notice as an almost essential step in the inductive process; and the more remarkable, as he is himself most fertile in technical phrases, always full of meaning, often happily expressed. Perhaps this omission, more than any other, obscured his full view of the progress of the sciences.

Mr. Herschel has devoted a chapter of his work to the subject of nomenclature, and has made many judicious observations upon its advantages and rules: perhaps, however, he has too much restricted his view to its use as shown in the process of classification, and the sciences of which this is the main ingredient. No one, indeed, can help being struck by the singular distinctness and facility with which botany manages her one hundred thousand species, or with the strange and helpless embarrassment in which mineralogy has contrived to involve the three hundred species which are all that she can produce. But the effects of a well-chosen nomenclature, or, at least, of a distinct terminology, extend as fully to the expression of laws as to the recognition of objects. Whether or no *polarization* and *depolarization* are good terms, some terms to express the facts which these recall have long been indispensable; and it would be in vain for any one, without substituting something better, to set about expelling these by attacking their etymological propriety. Every science which has made any advance at all, has more or fewer of such terms of art; and the more familiarly they come to be used, and the less technical the character they acquire, the more important are their functions and the steadier their influence.

[. . .]

Mr. Herschel's Second Part is devoted to the illustration of the rules by which true science must be conducted, and the type to which it must conform. Such a branch of philosophy, when systematized, might properly be called *inductive logic,* since it delivers the precepts of schemes of inductive inference, as the old syllogistic logic does of deductive. This portion of the treatise is a most valuable addition to our philosophical literature; for it is, we believe, the first attempt since

Bacon to deliver a connected body of rules of philosophizing which shall apply alike to the conduct of all researches directed to the discovery of laws of nature; it is consequently the first of all such attempts made to deliver such precepts by a person who had before him examples of their successful application. Extraordinary as is the prophetic sagacity with which Bacon has traced the course of his method, there are points of detail where his devices are inapplicable; and this was an inevitable result of his constructing his machinery without being able to set it in action. We have often wondered that some of those who claim so loudly to be looked upon as his scholars, have never thought of putting their researches in the technical forms which he prescribes. The second book of the Novum Organum contains a portion of his plan of a perfect scientific investigation, and it is extremely curious. We are to have, first, a 'comparentia ad intellectum omnium instantiarum notarum quæ in eâdem naturâ conveniunt,' a *muster* of all the relevant facts; next, a 'comparentia' or muster of the negative cases, a 'tabula declinationissive absentiæ in proximo;' then a 'tabula graduum,' a muster of the changes of intensity. Having these three tables, the work of induction begins by a *rejection* or *exclusive* process towards all conditions obviously not essential. This having been performed, be says, in due manner, 'there will remain, as it were, at the bottom (all the *volatile* opinions going off in smoke) an affirmative form or law, solid and true, and clearly defined.'

Though there is much that is right and philosophical in the spirit of this, it would be both difficult and frivolous to attempt, in cases of physical research, to comply with the formalities of such a process; and these directions of the master are noticeable principally as a curiosity. The next stage of Bacon's scheme is, however, both remarkable and instructive: it is the 'permissio intellectûs,' or step by which the mind is allowed to rush forwards, in order to snatch, if possible, some general conception by which the phenomena, already arranged, may be connected; this conception having to be limited and verified by a return to facts. Mr. Herschel fixes our attention on the same intellectual process, which Bacon had thus characterized: 'Such,' he observes,

> 'Such is the tendency of the human mind to speculation, that on the least idea of an analogy between a few phenomena, it leaps forward, as it were, to a cause or law, to the temporary neglect of all the rest; so that, in fact, almost all our principal inductions must be regarded as a series of ascents and descents, and of conclusions from a few cases, verified by trial on many.' – pp. 164, 165.

We shall not dwell longer on details. But there is one warning of the inductive legislator respecting the vicious indulgence of this spontaneous impulse, which it would be wrong not to mention. We refer to his condemnation of the method of *anticipation,* as opposed to that of gradual induction; a judgment indeed which of itself almost conveys the whole spirit and character of his philosophy, and which, therefore, we have been surprised to find not more distinctly touched upon in Mr.

Herschel's discourse. The way of anticipation *leaps at once* from facts and particulars to the most general axioms, and then, assuming these to be established, deduces from them the laws of particular cases.

> 'There are, and can be,' says Bacon, 'but these two ways of seeking truth; the former, the anticipatory, is the one now in use; the latter is the true but yet untried path.' (Aphorism 19.) 'Both begin from sense and from particulars, both rest ultimately in supreme generalizations, yet they are separated by an immense interval; for while one sets up at once certain abstract and useless generalities, the other rises by degrees, to general expressions of existing things.' (Aphorism 22.)

Against this fundamental mistake the teacher's voice is incessantly and urgently raised; and yet the error is far, even in our times, from being obsolete. Indeed some anticipatory process is to this day the course almost inevitably followed by all who have not been taught a wiser caution through the failures of others or their own. The usual history of physical theory has been the reign of one anticipation after another in unbroken succession. When the chemists found the four elements of earth, air, fire, and water no longer tenable, they immediately entrenched themselves in the *tria prima,* the three principles of salt, sulphur, and mercury; and in proportion as new experiments rebelled against such a representation, they tortured both facts and theory without mercy, rather than allow that they had attempted too hastily to dart upon the ultimate elements of bodies: and this turmoil of impetuous and devious error, accompanied by slowly advancing truth, has been incessantly urged on. Bacon's voice alone is heard above the confusion, proclaiming that such is not the true road; that anticipations may lean on some phenomena, may offer propositions of commanding generality, may produce assent, may supply materials for deduction; but that they are not interpretations of Nature, and do not enable us to command her services; that it is only through narrower propositions that the wider can be rightly reached; and that all the intellects of all ages might attempt in vain to erect a durable structure on the hollow ground of anticipation.[3]

In sciences which really have made any considerable progress, we have not much to fear the permanent establishment of errors arising from this source; for the necessity of verifying our generalizations, or of abandoning them, is irresistibly felt and acted on. Still, any laxity of criticism on this subject can tend only to foster a spirit of gratuitous theorizing, which will misemploy the cultivators of science, and mislead those who learn it through words alone. We will, on this account, notice Mr. Herschel's remark on the discovery of the atomic theory, and with the less scruple, because no one will suspect Mr. Herschel of considerately commending, or Mr. Dalton of pertinaciously pursuing, the faithless path of unverified anticipation. Mr. Herschel says of the law of definite proportions,

> 'The extreme simplicity which characterizes it, and which is itself an indication, not unequivocal, of its elevated rank in the scale of physical truths, had the effect of causing it to be announced at once by Mr. Dalton, in its most general terms, on the contemplation of a few instances, without passing through subordinate stages of painful inductive ascent by the intermedium of subordinate laws, such as, had the contrary course been pursued by him, would have been naturally preparatory to it, and such as would have led others to it by the prosecution of Wenzel's and Richter's researches, had they been duly attended to.' – pp. 305, 306.

Now, such language we cannot but think is liable to be mistaken. If Mr. Dalton had *guessed* the law to be true from a *few* instances, and done no more, he would have been the first person, whose name has been permanently connected with the history of science in virtue of such an unexamined simplification. It is easy to form such conjectures, and hardly more meritorious to be right than wrong in them. The simplicity of a doctrine is a most fallacious argument for its truth, and may as often retard as advance science. What could be more simple than the supposition that the heavenly bodies all moved in circles? Nothing can be more opposed to the notion of discoveries unbought by labour, than the persevering and unwearied trials, the patient eagerness, the gradual insight, of those who have earned the fame of true philosophers. The popular love of the striking and wonderful may indeed be caught by the idea of sudden and unprepared inspirations of the genius of discovery; but those who have acquired sounder views of the appointed course of true and permanent science, will share in no such feeling. They tell us, to borrow the language of another living philosopher of the university and of the school of Bacon,

> 'That the records of mankind offer no single instance of any great physical truth anticipated by mere guesses and conjectures; that philosophic wisdom consists in comprehending the last generalizations derived from facts, each of which is only known by experiment and observation; and in advancing, by such means, to those general laws by which all things are bound together: that inventive power in physics, unlike inventive power in works of art or of imagination, finds no employment in ideal creations, and only means the faculty by which the mind clearly apprehends the relations and analogies of things already known, and is thereby directed and urged on to the discovery of new facts by the help of new comparisons; that the history of ages has proved this way of slow and toilsome induction to be the only path which leads to physical truth.'[4]

We have not space to notice Mr. Herschel's admirable precepts and maxims for scientific research. We may, however, mention, as an instance of them, his ninth rule, where he observes, that in tracing the law of a compound effect, we sometimes subduct the effect of all the *known* causes, and thus leave as it were a *residual phenomenon* to be explained. In modern times, knowing already as we do the possible effect of so many causes, this process is peculiarly useful in disentangling from the complex of nature the portions which depend on laws still unknown. Mr. Herschel gives, as an example of this process, the return of Encke's comet, where, after allowing for all the perturbations produced by the attraction of the planets, there yet remains a retardation of the motion, seeming to indicate some material, though subtle medium, diffused through the celestial space: he instances it again in M. Arago's determination of the magnetic power of a copper plate, by subtracting from the retardation of a magnetic needle vibrating near it, the effect due to the known resistances of the surrounding air and the suspending thread; again, in the independent temperature of the celestial regions, which Fourrier conceived to remain after subtracting the influence of the sun and stars; and in the detection of new elements in the residual quantities of chemical researches, as in the case of lithia, discovered by Arfwedson. It is obvious that such a mode of discovery (new and curious as a subject of speculation) can only be employed when we have already discovered much. It is not till we have proceeded so far in the interpretation of Nature as to have learnt the greater part of her language, that we can thus make out the meaning of a few words still unknown, by considering their place in the context.

If we are to pay any deference to the spirit of our time and country, we must not quit our subject without saying a word on that portion of physical science which Bacon calls 'deductio ad praxim;' its application to the uses and needs of human life. That the consideration of this office of human knowledge gives a near interest and vivid reality to its speculations, no one probably will deny; and there is even a kind of dignity bestowed on the physical conditions of man's well being, when we see the highest efforts of which the human intellect is capable, the discoveries which it has required generations to mature, ministering to his wants and performing his bidding. Bacon in anticipation, and Mr. Herschel in retrospect, seem to love to dwell on the powers and pleasures which knowledge thus brings in her train. 'No doubt,' says the former, 'the sovereignty of man lieth hid in knowledge, wherein many things are reserved which kings with their treasure cannot buy, nor with their force command.'Mr. Herschel observes, with manifest delight, instances 'where the extremes of speculative refinement and of practical utility are brought to approximate.' Bacon points to the discovery of gunpowder, of silk, of the compass, of sugar, of paper, and of printing, and then asks if, when men were not inquiring, when, they were following other objects, inventions, so many and so useful fell in their way, as it were by chance, what may we look for when they are making such knowledge their object; pursuing it methodically and systematically, instead of by casual impulse and desultory courses? Our modern philosopher does not disdain to draw his arguments from the refinements which science has introduced into such arts as smelting, bleaching, soap-boiling, and sugar-boiling. The applications of iodine to the cure of the goître, of lemon-juice to the prevention of scurvy, of the magnetic mask to protect the lungs of the needle-maker, of the safety-lamp to the preservation of the miner, afford him a benevolent pleasure, in the spectacle of human miseries alleviated, and of the substantial benefits flowing to the ignorant from what might at first sight appear to be only fanciful and wanton amusements of the intellects of the idle.

We will give Mr. Herschel's account of a few of these cases.

> 'A soap-manufacturer remarks that the residuum of his ley, when exhausted of the alkali for which he employs it, produces a corrosion of his copper boiler for which he cannot account. He puts it into the hands of a scientific chemist for analysis, and the result is the discovery of one of the most singular and important chemical elements, iodine. The properties of this, being studied, are found to occur most appositely in illustration and support of a variety of new, curious, and instructive views then gaining ground in chemistry, and thus exercise a marked influence over the whole body of that science. Curiosity is excited: the origin of the new substance is traced to the sea-plants from whose ashes the principal ingredient of soap is obtained, and ultimately to the sea-water itself. It is thence hunted through nature, discovered in salt-mines and springs, and pursued into all bodies which have a marine origin; among the rest into sponge. A medical practitioner then calls to mind a reputed remedy for the cure of one of the most grievous and unsightly disorders to which the human species is subject – the *goître* – which infests the inhabitants of mountainous districts to an extent that in this favoured land we have happily no experience of, and which was said to have been originally cured by the ashes of burnt sponge. Led by this indication, he tries the effect of iodine on that complaint, and the result establishes the extraordinary fact that this singular substance, taken as a medicine, acts with the utmost promptitude and energy on *goître,* dissipating the largest and most inveterate in a short time, and acting (of course, like all medicines, even the most approved, with occasional failures) as a specific, or natural antagonist, against that odious deformity. It is thus that any accession to our knowledge of nature is sure, sooner or later, to make itself felt in some practical application, and that a benefit conferred on science by the casual observation or shrewd remark of even an unscientific or illiterate person infallibly repays itself with interest, though often in a way that could never have been at first contemplated.' – pp. 50–52.

> 'Who would have conceived that linen rags were capable of producing *more than their own weight* of sugar, by the simple agency of one of the cheapest and most abundant acids? – that dry bones could be a magazine of nutriment, capable of preservation for years, and ready to yield up their sustenance in the form best adapted to the support of life, on the application of that powerful agent, steam, which enters so largely into all our processes, or of an acid at once cheap and durable? – that sawdust itself is susceptible of conversion into a substance bearing no remote

> analogy to bread; and though certainly less palatable than that of flour, yet no way disagreeable, and both wholesome and digestible as well as highly nutritive?' – pp. 64, 65.

This is all most true and most valuable; and assuredly Bacon, and his disciple, Mr. Herschel, are not the philosophers whom we shall find, dwelling too exclusively on such considerations. But we conceive that it would be to take an unworthy and ignoble view of the claims of knowledge and the grounds of its value, if we were to derive its importance from its applications alone. For men engaged in remote and abstruse researches, we sometimes hear a plea offered, in a tone of apology, that no one can foresee the possible results of a discovery, and that the most recondite and abstract speculations have often come, by some strange and circuitous route, to have a bearing on the uses of daily life. This is rightly said, but this is not all. The true and worthy claim of knowledge is that which every lover of it feels in his own heart; – that it is valuable for its own sake; that truth is worth having because it is truth; the more worth having, the more pregnant and comprehensive the character which it possesses. Utilitarian moralists may maintain that we cannot have any wise motive of action except our own advantage; utilitarian philosophers may maintain that we cannot have any sufficient inducement for research except the tangible benefit of our expected discovery. The consciousness of every good man contradicts the former dogma; the irresistible impulse of every true philosopher – every man with the spirit of a discoverer – is inconsistent with the latter. Even if we were to confine ourselves to the pleasure produced, if we were to put the love of truth on a level with the love of turtle, still the former delicacy may probably be more widely and intensely relished, certainly more generally and equally diffused; and we do not see why the gratification which men may receive from knowing the laws which regulate the motions of light, is not as worthy our regard as that which they would derive from travelling from London to Brighton in an hour and a half, or from breakfasting on fresh strawberries every day in the year. But in fact the love of knowledge ought not to be degraded so far as to be weighed ounce for ounce against the pleasures of sense. It differs from them as a duty differs from an indulgence. Knowledge is followed because it is itself a good: it is an end, not a means. There are affections directed towards it as distinctly marked, and as elevated in their kind, as any other portion of our mental constitution. We are its suitors for itself, and not for its dower; and if we were allowed to borrow Mr. Shandy's favourite quotation, and to translate as freely as he was in the habit of doing, we, would say, 'Amicus Socrates,' utility is a pleasant companion, 'sed magis amica veritas,' but truth is a beloved friend.

We rejoice here to borrow the truly dignified and philosophical expressions of Mr. Herschel.

> 'The question "*cui bono*" to what practical end and advantage do your researches tend? is one which the speculative philosopher who loves knowledge for its own sake, and enjoys, as a rational being should enjoy,

> the mere contemplation of harmonious and mutually dependent truths, can seldom hear without a sense of humiliation. He feels that there is a lofty and disinterested pleasure in his speculations which ought to exempt them from such questioning; communicating as they do to his own mind the purest happiness (after the exercise of the benevolent and moral feelings) of which human nature is susceptible, and tending to the injury of no one, he might surely allege *this* as a sufficient and direct reply to those who, having themselves little capacity, and less relish for intellectual pursuits, are constantly repeating upon him this inquiry.' – p. 10.

If we had here space for the discussion, it would be a curious question to examine whether the attention of men of science may not be directed towards this practical deduction, this application of science to arts and commerce, so far as to interfere with their success in ascending to laws of a more exalted generality and higher speculative beauty; and we might ask whether such a diverse tendency of the scientific intellect of different countries has operated to produce any differences which may be detected in the present condition of England, as to the cultivation of science, compared with other countries: but we must hasten to conclude.

We have only a word or two to say on the style and manner of Mr. Herschel's treatise. Our men of science have not of late been considered remarkable for beauty or brilliancy of composition: the mathematicians and physical philosophers, especially, have often been reproached with having their style tinged with French expressions, arising from the influence of the many excellent works in that language relating to their studies. If such was the rule, however, there were splendid exceptions in Davy, Playfair, Leslie, and Brewster; and new names, worthy of being classed with these, are now daily multiplying among us. We may allude in particular to the strong and masculine eloquence with which Sedgewick announces his masterly and comprehensive views, and the clear, elegant, varied diction of Lyell. Mr. Herschel merits at least equal commendation for his sound, vigorous English, pregnant with thought and meaning, and very often marked with a well chosen pointedness of expression. Here and there, perhaps, we may detect a scientific Gallicism which has straggled in; as when he speaks of persons who have '*imagined* processes,' meaning *devised;* and of 'the *behaviour* of a particle set in motion when it meets another lying in its way' – a phrase which would probably startle the mere English reader, though it may not be easy for him to suggest an adequate substitute. But, on the other hand, we find scattered through the work instances of vivid and happy illustration, where the fancy is usefully called into action, so as sometimes to remind us of the splendid pictures which crowd upon us in the style of Bacon, where the restlessness of the intellect appears to keep its chamber-fellow, the imagination, perpetually awake.

We cannot better conclude our review of Mr. Herschel's delightful volume than by an extract from his first chapter, – one of the finest essays on the moral conduct of the intellect which has ever been produced.

'Nothing, then, can be more unfounded than the objection which has been taken, *in limine,* by persons, well meaning perhaps, certainly narrow-minded, against the study of natural philosophy, – that it fosters in its cultivators an undue and overweening self-conceit, leads them to doubt the immortality of the soul, and to scoff at revealed religion. Its natural effect, we may confidently assert, on every well constituted mind, is, and must be, the direct contrary. No doubt the testimony of natural reason, on whatever exercised, must of necessity stop short of those truths which it is the object of revelation to make known; but, while it places the existence and principal attributes of a Deity on such grounds as to render doubt absurd and atheism ridiculous, it unquestionably opposes no natural or necessary obstacle to further progress: on the contrary, by cherishing as a vital principle an unbounded spirit of inquiry and ardency of expectation, it unfetters the mind from prejudices of every kind, and leaves it open and free to every impression of a higher nature which it is susceptible of receiving, guarding only against enthusiasm and self-deception by a habit of strict investigation, but encouraging, rather than suppressing, everything that can offer a prospect or a hope beyond the present obscure and unsatisfactory state. The character of the true philosopher is to hope all things not impossible, and to believe all things not unreasonable. He who has seen obscurities which appeared impenetrable in physical and mathematical science suddenly dispelled, and the most barren and unpromising fields of inquiry converted, as if by inspiration, into rich and inexhaustible springs of knowledge and power, on a simple change of our point of view, or by merely bringing to bear on them some principle which it never occurred before to try, will surely be the very last to acquiesce in any dispiriting prospects of either the present or future destinies of mankind; while, on the other hand, the boundless views of intellectual and moral, as well as material, relations which open on him on all hands in the course of these pursuits, the knowledge of the trivial place he occupies in the scale of creation, and the sense continually pressed upon him of his own weakness and incapacity to suspend or modify the slightest movement of the vast machinery he sees in action around him, must effectually convince him that humility of pretension, no less than confidence of hope, is what best becomes his character.' – pp. 7, 8.

Notes

1 We may here notice what appears to us a want in the constitution of the English scientific world – the absence of any distinct and periodical call on any man of science to make such a survey of our intellectual acquisitions as we have referred to above. The valuable works of which we have spoken in the text are Reports given, in the one instance, to the French government and to the Institute, in the other, to the Academy of Sciences at Stockholm. The annual reports of such men as Cuvier on the advance of the natural

sciences, Berzelius on chemistry and physics, Fourrier on pure and applied mathematics, and the occasional reports of committees, such as are appointed by the Institute, are inevitably of great weight and of most beneficial efficacy. It is worth the while of an investigator to toil and think, when he is sure of having all that is really valuable in his results estimated at its true worth, connected with the history of science, and commended to the notice of scientific Europe by such men as these. We cannot imagine why Englishmen are not to present, and have presented to them, similar reports. We conceive that those who seek to add to the efficacy and dignity of our scientific bodies, could take no method so effectual towards their object, as this, of requiring from competent persons in appropriate situations annual reviews on science in general, or occasional reports on particular branches of it. They would thus find for the most abstruse communications of the cultivators of science 'fit audience, though few;' and the few might extend the communication, and, in some measure, the fitness, to a wider circle. The want of the habit of connecting what is discovered in this country with the general body of knowledge, and with the contemporary labours of foreigners, is one of the most powerful obstacles in the way of estimating the good at its true worth, and consequently of discountenancing what is trifling or erroneous.

2 Dr. Reid gave the name of the *inductive principle* to our instinctive belief in the permanent uniformity of the physical laws of nature. This is something very different from the inductive process by which we represent a number of particulars as classified under laws. Mr. Stewart (Phil. of the Human Mind, vol. ii., ch. iv., sec. 1,) is of opinion, that we are said to proceed according to the method of induction, 'when, by comparing a number of circumstances, agreeing in some cases, but differing in others, and all attended with the same results, the philosopher connects, as a general law of nature, the event with its *physical cause.*' But this is too narrow a view; for, as Mr. Herschel justly says (p. 9), 'dismissing, as beyond our reach, the inquiry into causes, we must be content at present to concentrate our attention on the *laws* which prevail among phenomena;' and the reduction of such laws, whether of causation or of any other connexion, according to an ascending scale of generality, he properly describes as induction.

It is hardly necessary to refer to the vague and popular, or as Bacon would call them, *plebeian* notions, which represent the Baconian method as residing in an appeal to phenomena, or in an analysis of phenomena, or in delivering knowledge in aphorisms, or in asserting the right of reason, or in anything which the declaimer may think fitted to elevate Bacon by comparison with his predecessors. A person must have read the Master very ill, if he has learnt to boast of the *obviousness* of the method, and to think scorn of the exoterics on that account.

Nor is it necessary to employ any space in distinguishing ours from the induction of mathematics, which is a kind of deduction beginning from a particular case; or from the meaning of the French word *induction,* which seems to be equivalent to inference.

3 Nov. Org., lib. i., Aph. 30.

4 President's address, at the anniversary of the Geological Society, 1830, by Professor Sedgwick.

Part 3

WILLIAM WHEWELL'S *HISTORY* AND *PHILOSOPHY OF THE INDUCTIVE SCIENCES*

Editorial Headnote

9. William Whewell, *The Philosophy of the Inductive Sciences*, 2nd edn (1847)

William Whewell (1794–1866) was perhaps most renowned in his time as a kind of commentator on science, as well as the master of Trinity College, Cambridge, possibly the most important academic post in Britain. Over the course of the 1830s and 1840s, and augmented by a number of revised and expanded editions, he published a five-volume exploration of science: first three volumes of the *History of the Inductive Sciences*, followed by two further volumes of *Philosophy of the Inductive Sciences*, from which these excerpts are drawn.

"Inductive" here must be read carefully, and in a much broader sense than that familiar to students of contemporary philosophy. Rather than only a form of argument proceeding from observations to a generalization, Whewell (like Herschel in the last selection) intended by "induction" to encompass the entire process of the generation, formulation, and testing of scientific knowledge, of which such arguments are an essential part, but only a part. What makes induction special is the fact that human ingenuity adds, over and above the observed facts, "some General Conception, which is given, not by the phenomena, but by the mind" (p. 133).

From the beginning of this excerpt, we notice an important and distinctive characteristic of Whewell's presentation of the philosophy of science: with a distinctly Kantian flair, he wonders whether some principles of Newton's mechanics might "contain their own evidence, and . . . not require proof from experiment" (p. 125). Let's consider his analysis of the principle of inertia as an example. One might, he writes, see it as an expression of the idea that any motion must have a cause, and several authors have thus maintained that we do not need to derive Newton's first law from experience – that is, that the first law might be an *a priori*, philosophical truth. But what argument or observation could tell us that there might not be some internal principle of all matter that motions are reduced by some constant amount per unit of time, causing all moving objects to steadily slow down, in violation of the first law? This, on the contrary, might lead us to think that experiment was required to teach us that inertia is constant over time.

Whewell resolves the question with a further piece of philosophical argumentation: time, he argues, is never understood "as a cause, but only as a condition of other causes producing their effects." A world in which bodies steadily and constantly slow down over time is impossible precisely because this would amount to time acting as an independent cause in that world, something that can never occur. For Whewell, then, while in fact the First Law was discovered experimentally, "we have now attained a point of view in which we see that it might have been certainly known to be true independently of experience." This, he argues, is an indication of the character of scientific methodology that "will hereafter be found to suggest very important views with regard both to the nature and prospects of our knowledge" (p. 128).

For Whewell, scientific knowledge therefore proceeds by an interplay between *a priori* philosophical axioms of causation, motion, matter, and so forth, and the *interpretation* of these axioms in particular empirical contexts, with regard to particular cases. It is important to see the extent of Whewell's divergence with Herschel on this score. While for Herschel hypotheses are always to be elaborated in a reciprocal movement with experimentation, and could always be corrected by further experience, Whewell's argument here amounts to the claim that we can arrive at knowledge of *a priori*, necessary truths about the external world. Once we have attained the "perspective" that lets us see how Newton's first law could be (or could have been) derived without reference to experience, we come to know that it is a necessary truth about the world, that it could not have been otherwise.

Having emphasized the role of both theorizing and *a priori* argumentation in science, Whewell is certainly no simplistic foe of speculation or conjecture. How, then, are we to carefully and accurately employ these methods? We must recognize that we regularly "detect the order and connexion which exist, by conceiving imaginary relations of order and connexion which have no existence" (p. 136), and therefore must be constantly on our guard to compare our hypotheses with observed experience. Whewell also emphasizes, as would be important to a number of later philosophers of science, that our hypotheses, invented to explain already known phenomena, should equally well predict unobserved, future phenomena (pp. 140–41). And, in perhaps his most celebrated contribution, he also argues for what he calls the *consilience of inductions*: when a hypothesis can predict not only future phenomena similar to those already known, but when "inductions from classes of facts altogether different have thus *jumped together*" (p. 141), this serves as a particularly strong form of confirmation for this hypothesis. Newtonian gravitation, for instance, explains not only the paths of the orbits of the planets, but also the procession of their equinoxes, an initially apparently unconnected fact about them; the wave theory of light, initially proposed to describe facts about refraction, also offers an understanding of complex phenomena of polarization. These theories even make room for such novel phenomena without needing what we would now call *ad hoc* modifications to their structure, in contrast with now discredited theories such as phlogistic chemistry or Ptolemaic astronomy.

9

'OF THE ESTABLISHMENT OF THE PRINCIPLES OF DYNAMICS' AND 'OF CERTAIN CHARACTERISTICS OF SCIENTIFIC INDUCTION' FROM *PHILOSOPHY OF THE INDUCTIVE SCIENCES*

William Whewell

Source: William Whewell, "Of the Establishment of the Principles of Dynamics," and "Of Certain Characteristics of Scientific Induction," from *The Philosophy of the Inductive Sciences, Founded Upon Their History*, 2 vols., 2nd edn (1847), pp. 1:215–1:227, 2:46–2:74.

CHAPTER VII.

OF THE ESTABLISHMENT OF THE PRINCIPLES OF DYNAMICS.

1. IN the History of Mechanics, I have traced the steps by which the three Laws of Motion and the other principles of mechanics were discovered, established, and extended to the widest generality of form and application. We have, in these laws, examples of principles which were, historically speaking, obtained by reference to experience. Bearing in mind the object and the result of the preceding discussions, we cannot but turn with much interest to examine these portions of science; to inquire whether there be any real difference in the grounds and nature between the knowledge thus obtained, and those truths which we have already contemplated; and which, as we have seen, contain their own evidence, and do not require proof from experiment.

2. *The First Law of Motion.* – The first law of motion is, that *When a body moves not acted upon by any force, it will go on perpetually in a straight line, and with a uniform velocity.* Now what is the real ground of our assent to this proposition? That it is not at first sight a self-evident truth, appears to be clear; since from the time of Aristotle to that of Galileo the opposite assertion was held to be true; and it was believed that all bodies in motion had, by their own nature, a constant tendency to move more and more slowly, so as to stop at last. This belief,

DOI: 10.4324/9781003263814-12

indeed, is probably even now entertained by most persons, till their attention is fixed upon the arguments by which the first law of motion is established. It is, however, not difficult to lead any person of a speculative habit of thought to see that the retardation which constantly takes place in the motion of all bodies when left to themselves, is, in reality, the effect of extraneous forces which destroy the velocity. A top ceases to spin because the friction against the ground and the resistance of the air gradually diminish its motion, and not because its motion has any internal principle of decay or fatigue. This may be shown, and was, in fact, shown by Hooke before the Royal Society, at the time when the laws of motion were still under discussion, by means of experiments in which the weight of the top is increased, and the resistance to motion offered by its support, is diminished; for by such contrivances its motion is made to continue much longer than it would otherwise do. And by experiments of this nature, although we can never remove the whole of the external impediments to continued motion, and although, consequently, there will always be some retardation; and an end of the motion of a body left to itself however long it may be delayed, must at last come; yet we can establish a conviction that if all resistance could be removed, there would be no diminution of velocity, and thus the motion would go on for ever.

If we call to mind the axioms which we formerly stated, as containing the most important conditions involved in the idea of Cause, it will be seen that our conviction in this case depends upon the first axiom of Causation, that nothing can happen without a cause. Every change in the velocity of the moving body must have a cause; and if the change can, in any manner, be referred to the presence of other bodies, these are said to exert *force* upon the moving body: and the conception of force is thus evolved from the general idea of cause. *Force is any cause which has motion, or change of motion, for its effect;* and thus, all the change of velocity of a body which can be referred to extraneous bodies, – as the air which surrounds it, or the support on which it rests, – is considered as the effect of forces; and this consideration is looked upon as explaining the difference between the motion which really takes place in the experiment, and that motion which, as the law asserts, would take place if the body were not acted on by any forces.

Thus the truth of the first law of motion depends upon the axiom that no change can take place without a cause; and follows from the definition of force, if we suppose that there can be none but an *external* cause of change. But in order to establish the law, it was necessary further to be assured that there is no *internal* cause of change of velocity belonging to all matter whatever, and operating in such a manner that the mere progress of time is sufficient to produce a diminution of velocity in all moving bodies. It appears from the history of mechanical science, that this latter step required a reference to observation and experiment; and that the first law of motion is so far, historically at least, dependent upon our experience.

But notwithstanding this historical evidence of the need which we have of a reference to observed facts, in order to place this first law of motion out of doubt, it has been maintained by very eminent mathematicians and philosophers, that the law is, in truth, evident of itself, and does not really rest upon experimental proof.

Such, for example, is the opinion of D'Alembert,[1] who offers what is called an *à priori* proof of this law; that is, a demonstration derived from our ideas alone. When a body is put in motion, either, he says, the cause which puts it in motion at first, suffices to make it move one foot, or the continued action of the cause during this foot is requisite for the motion. In the first case, the same reason which made the body proceed to the end of the first foot will hold for its going on through a second, a third, a fourth foot, and so on for any number. In the second case, the same reason which made the force continue to act during the first foot, will hold for its acting, and therefore for the body moving during each succeeding foot. And thus the body, once beginning to move, must go on moving for ever.

It is obvious that we might reply to this argument, that the reasons for the body proceeding during each succeeding foot may not necessarily be all the same; for among these reasons may be the time which has elapsed; and thus the velocity may undergo a change as the time proceeds: and we require observation to inform us that it does not do so.

Professor Playfair has presented nearly the same argument, although in a different and more mathematical form.[2] If the velocity change, says he, it must change according to some expression of calculation depending upon the time, or, in mathematical language, must be a *function* of the time. If the velocity diminish as the time increases, this may be expressed by stating the velocity in each case as a certain number, from which another quantity, or *term,* increasing as the time increases, is subtracted. But, Playfair adds, there is no condition involved in the nature of the case, by which the *coefficients,* or numbers which are to be employed, along with the number representing the time, in calculating this second term, can be determined to be of one magnitude rather than of any other. Therefore he infers there can be no such coefficients, and that the velocity is in each case equal to some constant number, independent of the time; and is therefore the same for all times.

In reply to this we may observe, that the circumstance of *our not seeing* in the nature of the case anything which determines for us the coefficients above spoken off, cannot prove that they have not some certain value *in nature.* We do not see in the nature of the case anything which should determine a body to fall sixteen feet in a second of time, rather than one foot or one hundred feet: yet in fact the space thus run through by falling bodies is determined to a certain magnitude. It would be easy to assign a mathematical expression for the velocity of a body, implying that one-hundredth of the velocity, or any other fraction, is lost in each second[3]: and where is the absurdity of supposing such an expression really to represent the velocity?

Most modern writers on mechanics have embraced the opposite opinion, and have ascribed our knowledge of this first law of motion to experience. Thus M. Poisson, one of the most eminent of the mathematicians who have written on this subject, says,[4] "We cannot affirm *à priori* that the velocity communicated to a body will not become slower and slower of itself, and end by being entirely extinguished. It is only by experience and induction that this question can be decided."

Yet it cannot be denied that there is much force in those arguments by which it is attempted to shew that the First Law of Motion, such as we find it, is more consonant to our conceptions than any other would be. The Law, as it exists, is the most simple that we can conceive. Instead of having to determine by experiments what is the law of the natural change of velocity, we find the Law to be that it does not change at all. To a certain extent, the Law depends upon the evident axiom, that no change can take place without a cause. But the question further occurs, whether the mere lapse of time may not be a cause of change of velocity. In order to ensure this, we have recourse to experiment; and the result is that time alone does not produce any such change. In addition to the conditions of change which we collect from our own Ideas, we ask of Experience what other conditions and circumstances she has to offer; and the answer is, that she can point out none. When we have removed the alterations which external causes, in our very conception of them, occasion, there are no longer any alterations. Instead of having to guide ourselves by experience, we learn that on this subject she has nothing to tell us. Instead of having to take into account a number of circumstances, we find that we have only to reject all circumstances. The velocity of a body remains unaltered by time alone, of whatever kind the body itself be.

But the doctrine that time alone is not a cause of change of velocity in any body is further recommended to us by this consideration; – that time is conceived by us not as a cause, but only as a condition of other causes producing their effects. Causes operate *in* time; but it is only when the cause exists, that the lapse of time can give rise to alterations. When therefore all external causes of change of velocity are supposed to be removed, the velocity must continue identical with itself, whatever the time which elapses. An eternity of negation can produce no positive result.

Thus, though the discovery of the First Law of Motion was made, historically speaking, by means of experiment, we have now attained a point of view in which we see that it might have been certainly known to be true independently of experience. This law in its ultimate form, when completely simplified and steadily contemplated, assumes the character of a self-evident truth. We shall find the same process to take place in other instances. And this feature in the progress of science will hereafter be found to suggest very important views with regard both to the nature and prospects of our knowledge.

3. *Gravity is a Uniform Force.* – We shall find observations of the same kind offering themselves in a manner more or less obvious, with regard to the other principles of Dynamics. The determination of the laws according to which bodies fall downwards by the common action of gravity, has already been noticed in the History of Mechanics,[5] as one of the earliest positive advances in the doctrine of motion. These laws were first rightly stated by Galileo, and established by reasoning and by experiment, not without dissent and controversy. The amount of these doctrines is this: That gravity is a uniform accelerating force; such a *uniform force* having this for its character, that it makes *the velocity increase in exact proportion to the time of motion.* The relation which the spaces described by the body bear

to the times in which they are described, is obtained by mathematical deduction from this definition of the force.

The clear Definition of a uniform accelerating force, and the Proposition that gravity is such a force, were co-ordinate and contemporary steps in this discovery. In defining accelerating force, reference, tacit or express, was necessarily made to the second of the general axioms respecting causation, – That causes are measured by their effects. Force, in the cases now under our notice, is conceived to be, as we have already stated, any cause which, acting from without, changes the motion of a body. It must, therefore, in this acceptation, be measured by the magnitude of the changes which are produced. But *in what manner* the changes of motion are to be employed as the measures of force, is learnt from observation of the facts which we see taking place in the world. Experience *interprets* the axiom of causation, from which otherwise we could not deduce any real knowledge. We may assume, in virtue of our general conceptions of force, that under the same circumstances, a greater change of motion implies a greater force producing it; but what are we to expect when the circumstances change? The weight of a body makes it fall from rest at first, and causes it to move more quickly as it descends lower. We may express this by saying, that gravity, the universal force which makes all terrestrial bodies fall when not supported, by its continuous action first *gives* velocity to the body when it has none, and afterwards *adds* velocity to that which the body already has. But how is the velocity added proportioned to the velocity which already exists? Force acting on a body at rest, and on a body in motion, appears under very different conditions; – how are the effects related? Let the force be conceived to be in both cases the *same,* since force is conceived to depend upon the extraneous bodies, and not upon the condition of the moving mass itself. But the force being the same, the effects may still be different. It is at first sight conceivable that the body, acted upon by the same gravity, may receive a less addition of velocity when it is already moving in the direction in which this gravity impels it; for if we ourselves push a body forwards, we can produce little additional effect upon it when it is already moving rapidly away from us. May it not be true, in like manner, that although gravity be always the same force, its effect depends upon the velocity which the body under its influence already possesses?

Observation and reasoning combined, as we have said, enabled Galileo to answer these questions. He asserted and proved that we may consistently and properly measure a force by the velocity which is by it generated in a body, in some certain time, as one second; and further, that if we adopt this measure, gravity will be a force of the same value under all circumstances of the body which it affects; since it appeared that, in fact, a falling body does receive equal increments of velocity in equal times from first to last.

If it be asked whether we could have known, anterior to, or independent of, experiment, that gravity is a uniform force in the sense thus imposed upon the term; it appears clear that we must reply, that we could not have attained to such knowledge, since other laws of the motion of bodies downwards are easily conceivable, and nothing but observation could inform us that one of these laws does not prevail in fact. Indeed, we may add, that the assertion that the force of gravity is uniform, is so far from being self-evident, that it is not even true; for gravity varies according to the distance from the center of the earth; and although this variation is so small as to be, in the case of falling bodies, imperceptible, it negatives the rigorous uniformity of the force as completely, though not to the same extent, as if the weight of a body diminished in a marked degree, when it was carried from the lower to the upper room of a house. It cannot, then, be a truth independent of experience, that gravity is uniform.

Yet, in fact, the assertion that gravity is uniform was assented to, not only before it was proved, but even before it was clearly understood. It was readily granted by all, that bodies which fall freely are *uniformly* accelerated; but while some held the opinion just stated, that uniformly accelerated motion is that in which the velocity increases in proportion to the *time,* others maintained, that *that* is uniformly accelerated motion, in which the velocity increases in proportion to the *space;* so that, for example, a body in falling vertically through twenty feet should acquire twice as great a velocity as one which falls through ten feet.

These two opinions are both put forward by the interlocutors of Galileo's Dialogue on this subject.[6] And the latter supposition is rejected, the author showing, not that it is inconsistent with experience, but that it is impossible in itself: inasmuch as it would inevitably lead to the conclusion, that the fall through a large and a small vertical space would occupy exactly the same time.

Indeed, Galileo assumes his definition of uniformly accelerated motion as one which is sufficiently recommended by its own simplicity. "If we attend carefully," he says, "we shall find that no mode of increase of velocity is more simple than that which adds equal increments in equal times. Which we may easily understand if we consider the close affinity of time and motion: for as the uniformity of motion is defined by the equality of spaces described in equal times, so we may conceive the uniformity of acceleration to exist when equal velocities are added in equal times."

Galileo's mode of supporting his opinion, that bodies falling by the action of gravity are thus uniformly accelerated, consists, in the first place, in adducing the maxim that nature always employs the most simple means.[7] But he is far from considering this a decisive argument. "I," says one of his speakers, "as it would be very unreasonable in me to gainsay this or any other definition which any author may please to make, since they are all arbitrary, may still, without offence, doubt whether such a definition, conceived and admitted in the abstract, fits, agrees, and is verified in that kind of accelerated motion which bodies have when they descend naturally."

The experimental proof that bodies, when they fall downwards, are uniformly accelerated, is (by Galileo) derived from the inclined plane; and therefore assumes the proposition, that if such uniform acceleration prevail in vertical motion, it will also hold when a body is compelled to describe an oblique rectilinear path. This proposition may be shown to be true, if (assuming by anticipation the Third Law of Motion, of which we shall shortly have to speak,) we introduce the conception of a uniform statical force as the cause of uniform acceleration. For the force on the inclined plane bears a constant proportion to the vertical force, and this proportion is known from statical considerations. But in the work of which we are speaking, Galileo does not introduce this abstract conception of force as the foundation of his doctrines. Instead of this, he proposes, as a postulate sufficiently evident to be made the basis of his reasonings, That bodies which descend down inclined planes of different inclinations, but of the same vertical height, all acquire the same velocity.[8] But when this postulate has been propounded by one of the persons of the dialogue, another interlocutor says, "You discourse very probably; but besides this likelihood, I wish to augment the probability so far, that it shall be almost as complete as a necessary demonstration." He then proceeds to describe a very ingenious and simple experiment, which shows that when a body is made to swing upwards at the end of a string, it attains to the same height, whatever is the path it follows, so long as it starts from the lowest point with the same velocity. And thus Galileo's postulate is experimentally confirmed, so far as the force of gravity can be taken as an example of the forces which the postulate contemplates: and conversely, gravity is proved to be a uniform force, so far as it can be considered clear that the postulate is true of uniform forces.

When we have introduced the conception and definition of accelerating force, Galileo's postulate, that bodies descending down inclined planes of the same vertical height, acquire the same velocity, may, by a few steps of reasoning, be demonstrated to be true of uniform forces: and thus the proof that gravity, either in vertical or oblique motion, is a uniform force, is confirmed by the experiment above mentioned; as it also is, on like grounds, by many other experiments, made upon inclined planes and pendulums.

Thus the propriety of Galileo's conception of a uniform force, and the doctrine that gravity is a uniform force, were confirmed by the same reasonings and experiments. We may make here two remarks; *First,* that the conception, when established and rightly stated, appears so simple as hardly to require experimental proof; a remark which we have already made with regard to the First Law of Motion: and *Second,* that the discovery of the real law of nature was made by assuming propositions which, without further proof, we should consider as very precarious, and as far less obvious, as well as less evident, than the law of nature in its simple form.

[. . .]

CHAPTER V.

OF CERTAIN CHARACTERISTICS OF SCIENTIFIC INDUCTION.

SECT. I. – *Invention a part of Induction.*

1. THE two operations spoken of in the preceding chapters, – the Explication of the Conceptions of our own minds, and the Colligation of observed Facts by

the aid of such Conceptions, – are, as we have just said, inseparably connected with each other. When united, and employed in collecting knowledge from the phenomena which the world presents to us, they constitute the mental process of *Induction;* which is usually and justly spoken of as the genuine source of all our *real general knowledge* respecting the external world. And we see, from the preceding analysis of this process into its two constituents, from what origin it derives each of its characters. It is *real,* because it arises from the combination of Real Facts, but it is *general,* because it implies the possession of General Ideas. Without the former, it would not be knowledge of the External World; without the latter, it would not be Knowledge at all. When Ideas and Facts are separated from each other, the neglect of Facts gives rise to empty speculations, idle subtleties, visionary inventions, false opinions concerning the laws of phenomena, disregard of the true aspect of nature: while the want of Ideas leaves the mind overwhelmed, bewildered, and stupified by particular sensations, with no means of connecting the past with the future, the absent with the present, the example with the rule; open to the impression of all appearances, but capable of appropriating none. Ideas are the *Form,* facts the *Material,* of our structure. Knowledge does not consist in the empty mould, or in the brute mass of matter, but in the rightly-moulded substance. Induction gathers general truths from particular facts; – and in her harvest, the corn and the reaper, the solid ears and the binding band, are alike requisite. All our knowledge of nature is obtained by Induction; the term being understood according to the explanation we have now given. And our knowledge is then most complete, then most truly deserves the name of Science, when both its elements are most perfect; – when the Ideas which have been concerned in its formation have, at every step, been clear and consistent; – and when they have, at every step also, been employed in binding together real and certain Facts. Of such Induction, I have already given so many examples and illustrations in the two preceding chapters, that I need not now dwell further upon the subject.

2. Induction is familiarly spoken of as the process by which we collect a *General Proposition* from a number of *Particular Cases:* and it appears to be frequently imagined that the general proposition results from a mere juxta-position of the cases, or at most, from merely conjoining and extending them. But if we consider the process more closely, as exhibited in the cases lately spoken of, we shall perceive that this is an inadequate account of the matter. The particular facts are not merely brought together, but there is a New Element added to the combination by the very act of thought by which they are combined. There is a Conception of the mind introduced in the general proposition, which did not exist in any of the observed facts. When the Greeks, after long observing the motions of the planets, saw that these motions might be rightly considered as produced by the motion of one wheel revolving in the inside of another wheel, these Wheels were Creations of their minds, added to the Facts which they perceived by sense. And even if the wheels were no longer supposed to be material, but were reduced to mere geometrical spheres or circles, they were not the less products of the mind alone, – something additional to the facts observed. The same is the case in all other discoveries.

The facts are known, but they are insulated and unconnected, till the discoverer supplies from his own stores a Principle of Connexion. The pearls are there, but they will not hang together till some one provides the String. The distances and periods of the planets were all so many separate facts; by Kepler's Third Law they are connected into a single truth: but the Conceptions which this law involves were supplied by Kepler's mind, and without these, the facts were of no avail. The planets described ellipses round the sun, in the contemplation of others as well as of Newton; but Newton conceived the deflection from the tangent in these elliptical motions in a new light, – as the effect of a Central Force following a certain law; and then it was, that such a force was discovered truly to exist.

Thus[9] in each inference made by Induction, there is introduced some General Conception, which is given, not by the phenomena, but by the mind. The conclusion is not contained in the premises, but includes them by the introduction of a New Generality. In order to obtain our inference, we travel beyond the cases which we have before us; we consider them as mere exemplifications of some Ideal Case in which the relations are complete and intelligible. We take a Standard, and measure the facts by it; and this Standard is constructed by us, not offered by Nature. We assert, for example, that a body left to itself will move on with unaltered velocity; not because our senses ever disclosed to us a body doing this, but because (taking this as our Ideal Case) we find that all actual cases are intelligible and explicable by means of the Conception of *Forces,* causing change and motion, and exerted by surrounding bodies. In like manner, we see bodies striking each other, and thus moving and stopping, accelerating and retarding each other: but in all this, we do not perceive by our senses that abstract quantity, *Momentum,* which is always lost by one body as it is gained by another. This Momentum is a creation of the mind, brought in among the facts, in order to convert their apparent confusion into order, their seeming chance into certainty, their perplexing variety into simplicity. This the Conception of *Momentum gained and lost* does: and in like manner, in any other case in which a truth is established by Induction, some Conception is introduced, some Idea is applied, as the means of binding together the facts, and thus producing the truth.

3. Hence in every inference by Induction, there is some Conception *superinduced* upon the Facts: and we may henceforth conceive this to be the peculiar import of the term *Induction.* I am not to be understood as asserting that the term was originally or anciently employed with this notion of its meaning; for the peculiar feature just pointed out in Induction has generally been over-looked. This appears by the accounts generally given of Induction. "Induction," says Aristotle,[10] "is when by means of one extreme term[11] we infer the other extreme term to be true of the middle term." Thus, (to take such exemplifications as belong to our subject,) from knowing that Mercury, Venus, Mars, describe ellipses about the Sun, we infer that all Planets describe ellipses about the Sun. In making this inference syllogistically, we assume that the evident proposition, "Mercury, Venus, Mars, do what all Planets do," may be taken *conversely,* "All Planets do what Mercury, Venus, Mars, do." But we may remark that, in this passage, Aristotle (as was natural in his line

of discussion) turns his attention entirely to the *evidence* of the inference; and overlooks a step which is of far more importance to our knowledge, namely, the *invention* of the second extreme term. In the above instance, the particular luminaries, Mercury, Venus, Mars, are one logical *Extreme;* the general designation Planets is the *Middle Term;* but having these before us, how do we come to think of *description of ellipses,* which is the other Extreme of the syllogism? When we have once invented this "second Extreme Term," we may, or may not, be satisfied with the evidence of the syllogism; we may, or may not, be convinced that, so far as this property goes, the extremes are co-extensive with the middle term[12]; but the *statement* of the syllogism is the important step in science. We know how long Kepler laboured, how hard he fought, how many devices he tried, before he hit upon this *Term,* the Elliptical Motion. He rejected, as we know, many other "second Extreme Terms," for example, various combinations of epicyclical constructions, because they did not represent with sufficient accuracy the special facts of observation. When he had established his premiss, that "Mars does describe an Ellipse about the Sun," he does not hesitate to *guess* at least that, in this respect, he might *convert* the other premiss, and assert that "All the Planets do what Mars does." But the main business was, the inventing and verifying the proposition respecting the Ellipse. The Invention of the Conception was the great step in the *discovery;* the Verification of the Proposition was the great step in the *proof* of the discovery. If Logic consists in pointing out the conditions of proof, the Logic of Induction must consist in showing what are the conditions of proof, in such inferences as this: but this subject must be pursued in the next chapter; I now speak principally of the act of *Invention,* which is requisite in every inductive inference.

4. Although in every inductive inference, an act of invention is requisite, the act soon slips out of notice. Although we bind together facts by superinducing upon them a new Conception, this Conception, once introduced and applied, is looked upon as inseparably connected with the facts, and necessarily implied in them. Having once had the phenomena bound together in their minds in virtue of the Conception, men can no longer easily restore them back to the detached and incoherent condition in which they were before they were thus combined. The pearls once strung, they seem to form a chain by their nature. Induction has given them a unity which it is so far from costing us an effort to preserve, that it requires an effort to imagine it dissolved. For instance, we usually represent to ourselves the Earth as *round,* the Earth and the Planets as *revolving* about the Sun, and as *drawn* to the Sun by a Central Force; we can hardly understand how it could cost the Greeks, and Copernicus, and Newton, so much pains and trouble to arrive at a view which is to us so familiar. These are no longer to us Conceptions caught hold of and kept hold of by a severe struggle; they are the simplest modes of conceiving the facts: they are really Facts. We are willing to *own* our obligation to those discoverers, but we hardly *feel* it: for in what other manner (we ask in our thoughts,) could we represent the facts to ourselves?

Thus we see why it is that this step of which we now speak, the Invention of a new Conception in every inductive inference, is so generally overlooked that it has hardly been noticed by preceding philosophers. When once performed by the discoverer, it takes a fixed and permanent place in the understanding of every one. It is a thought which, once breathed forth, permeates all men's minds. All fancy they nearly or quite knew it before. It oft was thought, or almost thought, though never till now expressed. Men accept it and retain it, and know it cannot be taken from them, and look upon it as their own. They will not and cannot part with it, even though they may deem it trivial and obvious. It is a secret, which once uttered, cannot be recalled, even though it be despised by those to whom it is imparted. As soon as the leading term of a new theory has been pronounced and understood, all the phenomena change their aspect. There is a standard to which we cannot help referring them. We cannot fall back into the helpless and bewildered state in which we gazed at them when we possessed no principle which gave them unity. Eclipses arrive in mysterious confusion: the notion of a *Cycle* dispels the mystery. The Planets perform a tangled and mazy dance; but *Epicycles* reduce the maze to order. The Epicycles themselves run into confusion; the conception of an *Ellipse* makes all clear and simple. And thus from stage to stage, new elements of intelligible order are introduced. But this intelligible order is so completely adopted by the human understanding, as to seem part of its texture. Men ask Whether Eclipses follow a Cycle; Whether the Planets describe Ellipses; and they imagine that so long as they do not *answer* such questions rashly, they take nothing for granted. They do not recollect how much they assume in *asking* the question: – how far the conceptions of Cycles and of Ellipses are beyond the visible surface of the celestial phenomena: – how many ages elapsed, how much thought, how much observation, were needed, before men's thoughts were fashioned into the words which they now so familiarly use. And thus they treat the subject, as we have seen Aristotle treating it; as if it were a question, not of invention, but of proof; not of substance, but of form: as if the main thing were not *what* we assert, but *how* we assert it. But for our purpose, it is requisite to bear in mind the feature which we have thus attempted to mark; and to recollect that, in every inference by induction, there is a Conception supplied by the mind and superinduced upon the Facts.

5. In collecting scientific truths by Induction, we often find (as has already been observed), a Definition and a Proposition established at the same time, – introduced together, and mutually dependent on each other. The combination of the two constitutes the Inductive act; and we may consider the Definition as representing the superinduced Conception, and the Proposition as exhibiting the Colligation of Facts.

Sect. II. – *Use of Hypotheses.*

6. To discover a Conception of the mind which will justly represent a train of observed facts is, in some measure, a process of conjecture, as I have stated already; and as I then observed, the business of conjecture is commonly conducted by calling up before our minds several suppositions, and selecting that one which most agrees with what we know of the observed facts. Hence he who

has to discover the laws of nature may have to invent many suppositions before he hits upon the right one; and among the endowments which lead to his success, we must reckon that fertility of invention which ministers to him such imaginary schemes, till at last he finds the one which conforms to the true order of nature. A facility in devising hypotheses, therefore, is so far from being a fault in the intellectual character of a discoverer, that it is, in truth, a faculty indispensable to his task. It is, for his purposes, much better that he should be too ready in contriving, too eager in pursuing systems which promise to introduce law and order among a mass of unarranged facts, than that he should be barren of such inventions and hopeless of such success. Accordingly, as we have already noticed, great discoverers have often invented hypotheses which would not answer to all the facts, as well as those which would; and have fancied themselves to have discovered laws, which a more careful examination of the facts overturned.

The tendencies of our speculative nature,[13] carrying us onwards in pursuit of symmetry and rule, and thus producing all true theories, perpetually show their vigour by overshooting the mark. They obtain something, by aiming at much more. They detect the order and connexion which exist, by conceiving imaginary relations of order and connexion which have no existence. Real discoveries are thus mixed with baseless assumptions; profound sagacity is combined with fanciful conjecture; not rarely, or in peculiar instances, but commonly, and in most cases; probably in all, if we could read the thoughts of discoverers as we read the books of Kepler. To try wrong guesses is, with most persons, the only way to hit upon right ones. The character of the true philosopher is, not that he never conjectures hazardously, but that his conjectures are clearly conceived, and brought into rigid contact with facts. He sees and compares distinctly the Ideas and the Things; – the relations of his notions to each other and to phenomena. Under these conditions, it is not only excusable, but necessary for him, to snatch at every semblance of general rule, – to try all promising forms of simplicity and symmetry.

Hence advances in knowledge[14] are not commonly made without the previous exercise of some boldness and license in guessing. The discovery of new truths requires, undoubtedly, minds careful and scrupulous in examining what is suggested; but it requires, no less, such as are quick and fertile in suggesting. What is Invention, except the talent of rapidly calling before us the many possibilities, and selecting the appropriate one? It is true, that when we have rejected all the inadmissible suppositions, they are often quickly forgotten; and few think it necessary to dwell on these discarded hypotheses, and on the process by which they were condemned. But all who discover truths, must have reasoned upon many errours to obtain each truth; every accepted doctrine must have been one chosen out of many candidates. If many of the guesses of philosophers of bygone times now appear fanciful and absurd, because time and observation have refuted them, others, which were at the time equally gratuitous, have been confirmed in a manner which makes them appear marvellously sagacious. To form hypotheses, and then to employ much labour and skill in refuting, if they do not succeed in establishing them, is a part of the usual process of inventive minds. Such a proceeding belongs to the *rule* of the genius of discovery, rather than (as has often been taught in modern times) to the *exception.*

7. But if it be an advantage for the discoverer of truth that he be ingenious and fertile in inventing hypotheses which may connect the phenomena of nature, it is indispensably requisite that he be diligent and careful in comparing his hypotheses with the facts, and ready to abandon his invention as soon as it appears that it does not agree with the course of actual occurrences. This constant comparison of his own conceptions and supposition with observed facts under all aspects, forms the leading employment of the discoverer: this candid and simple love of truth, which makes him willing to suppress the most favourite production of his own ingenuity as soon as it appears to be at variance with realities, constitutes the first characteristic of his temper. He must have neither the blindness which cannot, nor the obstinacy which will not, perceive the discrepancy of his fancies and his facts. He must allow no indolence, or partial views, or self-complacency, or delight in seeming demonstration, to make him tenacious of the schemes which he devises, any further than they are confirmed by their accordance with nature. The framing of hypotheses is, for the inquirer after truth, not the end, but the beginning of his work. Each of his systems is invented, not that he may admire it and follow it into all its consistent consequences, but that he may make it the occasion of a course of active experiment and observation. And if the results of this process contradict his fundamental assumptions, however ingenious, however symmetrical, however elegant his system may be, he rejects it without hesitation. He allows no natural yearning for the offspring of his own mind to draw him aside from the higher duty of loyalty to his sovereign, Truth: to her he not only gives his affections and his wishes, but strenuous labour and scrupulous minuteness of attention.

We may refer to what we have said of Kepler, Newton, and other eminent philosophers, for illustrations of this character. In Kepler we have remarked[15] the courage and perseverance with which he undertook and executed the task of computing his own hypotheses: and, as a still more admirable characteristic, that he never allowed the labour he had spent upon any conjecture to produce any reluctance in abandoning the hypothesis, as soon as he had evidence of its inaccuracy. And in the history of Newton's discovery that the moon is retained in her orbit by the force of gravity, we have noticed the same moderation in maintaining the hypothesis, after it had once occurred to the author's mind. The hypothesis required that the moon should fall from the tangent of her orbit every second through a space of sixteen feet; but according to his first calculations it appeared that in fact she only fell through a space of thirteen feet in that time. The difference seems small, the approximation encouraging, the theory plausible; a man in love with his own fancies would readily have discovered or invented some probable cause of the difference. But Newton acquiesced in it as a disproof of his conjecture, and "laid aside at that time any further thoughts of this matter."[16]

8. It has often happened that those who have undertaken to instruct mankind have not possessed this pure love of truth and comparative indifference to the

maintenance of their own inventions. Men have frequently adhered with great tenacity and vehemence to the hypotheses which they have once framed; and in their affection for these, have been prone to overlook, to distort, and to misinterpret facts. In this manner, *Hypotheses* have so often been prejudicial to the genuine pursuit of truth, that they have fallen into a kind of obloquy; and have been considered as dangerous temptations and fallacious guides. Many warnings have been uttered against the fabrication of hypotheses by those who profess to teach philosophy; many disclaimers of such a course by those who cultivate science.

Thus we shall find Bacon frequently discommending this habit, under the name of "anticipation of the mind," and Newton thinks it necessary to say emphatically "hypotheses non fingo." It has been constantly urged that the inductions by which sciences are formed must be *cautious* and *rigorous;* and the various imaginations which passed through Kepler's brain, and to which he has given utterance, have been blamed or pitied as lamentable instances of an unphilosophical frame of mind. Yet it has appeared in the preceding remarks that hypotheses rightly used are among the helps, far more than the dangers, of science; – that scientific induction is not a "cautious" or a "rigorous" process in the sense of *abstaining from* such suppositions, but in *not adhering to* them till they are confirmed by fact, and in carefully seeking from facts confirmation or refutation. Kepler's character was, not that he was peculiarly given to the construction of hypotheses, but that he narrated with extraordinary copiousness and candour the course of his thoughts, his labours, and his feelings. In the minds of most persons, as we have said, the inadmissible suppositions, when rejected, are soon forgotten: and thus the trace of them vanishes from the thoughts, and the successful hypothesis alone holds its place in our memory. But in reality, many other transient suppositions must have been made by all discoverers; – hypotheses which are not afterwards asserted as true systems, but entertained for an instant; – "tentative hypotheses," as they have been called. Each of these hypotheses is followed by its corresponding train of observations, from which it derives its power of leading to truth. The hypothesis is like the captain, and the observations like the soldiers of an army: while he appears to command them, and in this way to work his own will, he does in fact derive all his power of conquest from their obedience, and becomes helpless and useless if they mutiny.

Since the discoverer has thus constantly to work his way onwards by means of hypotheses, false and true, it is highly important for him to possess talents and means for rapidly *testing* each supposition as it offers itself. In this as in other parts of the work of discovery, success has in general been mainly owing to the native ingenuity and sagacity of the discoverer's mind. Yet some Rules tending to further this object have been delivered by eminent philosophers, and some others may perhaps be suggested. Of these we shall here notice only some of the most general, leaving for a future chapter the consideration of some more limited and detailed processes by which, in certain cases, the discovery of the laws of nature may be materially assisted.

Sect. III. – *Tests of Hypotheses.*

9. A Maxim which it may be useful to recollect is this; – that *hypotheses may often be of service to science, when they involve a certain portion of incompleteness, and even of errour.* The object of such inventions is to bind together facts which without them are loose and detached; and if they do this, they may lead the way to a perception of the true rule by which the phenomena are associated together, even if they themselves somewhat misstate the matter. The imagined arrangement enables us to contemplate, as a whole, a collection of special cases which perplex and overload our minds when they are considered in succession; and if our scheme has so much of truth in it as to conjoin what is really connected, we may afterwards duly correct or limit the mechanism of this connexion. If our hypothesis renders a reason for the agreement of cases really similar, we may afterwards find this reason to be false, but we shall be able to translate it into the language of truth.

A conspicuous example of such an hypothesis, one which was of the highest value to science, though very incomplete, and as a representation of nature altogether false, is seen in the *Doctrine of epicycles* by which the ancient astronomers explained the motions of the sun, moon, and planets. This doctrine connected the places and velocities of these bodies at particular times in a manner which was, in its general features, agreeable to nature. Yet this doctrine was erroneous in its assertion of the circular nature of all the celestial motions, and in making the heavenly bodies revolve round the earth. It was, however, of immense value to the progress of astronomical science; for it enabled men to express and reason upon many important truths which they discovered respecting the motion of the stars, up to the time of Kepler. Indeed we can hardly imagine that astronomy could, in its outset, have made so great a progress under any other form, as it did in consequence of being cultivated in this shape of the incomplete and false epicyclical hypothesis.

We may notice another instance of an exploded hypothesis, which is generally mentioned only to be ridiculed, and which undoubtedly is both false in the extent of its assertion, and unphilosophical in its expression; but which still, in its day, was not without merit. I mean the doctrine of *Nature's horrour of a vacuum* (*fuga vacui*), by which the action of siphons and pumps and many other phenomena were explained, till Mersenne and Pascal taught a truer doctrine. This hypothesis was of real service; for it brought together many facts which really belong to the same class, although they are very different in their first aspect. A scientific writer of modern times[17] appears to wonder that men did not at once divine the weight of the air from which the phenomena formerly ascribed to the *fuga vacui* really result. "Loaded, compressed by the atmosphere," he says, "they did not recognize its action. In vain all nature testified that air was elastic and heavy; they shut their eyes to her testimony. The water rose in pumps and flowed in siphons at that time, as it does at this day. They could not separate the boards of a pair of bellows of which the holes were stopped; and they could not bring together the same boards without difficulty, if they were at first separated. Infants sucked the milk of their mothers; air entered rapidly into the lungs of animals at every inspiration; cupping-glasses produced tumours on the skin; and in spite of all the striking proofs of the weight and elasticity of the air, the ancient philosophers maintained resolutely that air was light, and explained all these phenomena by the horrour which they said nature had for a vacuum." It is curious that it should not have occurred to the author while writing this, that if these facts, so numerous and various, can all be accounted for by *one* principle, there is a strong presumption that the principle is not altogether baseless. And in reality is it not true that nature *does* abhor a vacuum, and do all she can to avoid it? No doubt this power is not unlimited; and we can trace it to a mechanical cause, the pressure of the circumambient air. But the tendency, arising from this pressure, which the bodies surrounding a space void of air have to rush into it, may be expressed, in no extravagant or unintelligible manner, by saying that nature has a repugnance to a vacuum.

That imperfect and false hypotheses, though they may thus explain *some* phenomena, and may be useful in the progress of science, cannot explain *all* phenomena; – and that we are never to rest in our labours or acquiesce in our results, till we have found some view of the subject which *is* consistent with *all* the observed facts: – will of course be understood. We shall afterwards have to speak of the other steps of such a progress.

10. The hypotheses which we accept ought to explain phenomena which we have observed. But they ought to do more than this: our hypotheses ought to *foretel* phenomena which have not yet been observed; – at least all of the same kind as those which the hypothesis was invented to explain. For our assent to the hypothesis implies that it is held to be true of all particular instances. That these cases belong to past or to future times, that they have or have not already occurred, makes no difference in the applicability of the rule to them. Because the rule prevails, it includes all cases; and will determine them all, if we can only calculate its real consequences. Hence it will predict the results of new combinations, as well as explain the appearances which have occurred in old ones. And that it does this with certainty and correctness, is one mode in which the hypothesis is to be verified as right and useful.

The scientific doctrines which have at various periods been established have been verified in this manner. For example, the *Epicyclical Theory* of the heavens was confirmed by its *predicting* truly eclipses of the sun and moon, configurations of the planets, and other celestial phenomena; and by its leading to the construction of Tables by which the places of the heavenly bodies were given at every moment of time. The truth and accuracy of these predictions were a proof that the hypothesis was valuable and, at least to a great extent, true; although, as was afterwards found, it involved a false representation of the structure of the heavens. In like manner, the discovery of the *Laws of Refraction* enabled mathematicians to *predict,* by calculation, what would be the effect of any new form or combination of transparent lenses. Newton's hypothesis of *Fits of Easy Transmission and Easy Reflection* in the particles of light, although not confirmed by other kinds of facts, involved a true statement of the law of the phenomena which it was framed to include, and served to *predict* the forms and colours of thin plates for a wide range of given cases. The hypothesis that Light operates by *Undulations* and *Interferences,* afforded the means of *predicting* results under a still larger extent of conditions. In like manner in the progress of chemical knowledge, the doctrine of *Phlogiston* supplied the means of *foreseeing* the consequence of many combinations of elements, even before they were tried; but the *Oxygen Theory,* besides affording predictions, at least equally exact, with regard to the general results of chemical operations, included all the facts concerning the relations of weight of the elements and their compounds, and enabled chemists to *foresee* such facts in untried cases. And the Theory of *Electromagnetic Forces,* as soon as it was rightly understood, enabled those who had mastered it to *predict* motions such as had not been before observed, which were accordingly found to take place.

Men cannot help believing that the laws laid down by discoverers must be in a great measure identical with the real laws of nature, when the discoverers thus determine effects beforehand in the same manner in which nature herself determines them when the occasion occurs. Those who can do this, must, to a considerable extent, have detected nature's secret; – must have fixed upon the conditions to which she attends, and must have seized the rules by which she applies them. Such a coincidence of untried facts with speculative assertions cannot be the work of chance, but implies some large portion of truth in the principles on which the reasoning is founded. To trace order and law in that which has been observed, may be considered as interpreting what nature has written down for us, and will commonly prove that we understand her alphabet. But to predict what has not been observed, is to attempt ourselves to use the legislative phrases of nature; and when she responds plainly and precisely to that which we thus utter, we cannot but suppose that we have in a great measure made ourselves masters of the meaning and structure of her language. The prediction of results, even of the same kind as those which have been observed, in new cases, is a proof of real success in our inductive processes.

11. We have here spoken of the prediction of facts *of the same kind* as those from which our rule was collected. But the evidence in favour of our induction is of a much higher and more forcible character when it enables us to explain and determine cases of a *kind different* from those which were contemplated in the formation of our hypothesis. The instances in which this has occurred, indeed, impress us with a conviction that the truth of our hypothesis is certain. No accident could give rise to such an extraordinary coincidence. No false supposition could, after being adjusted to one class of phenomena, exactly represent a different class, when the agreement was unforeseen and uncontemplated. That rules springing from remote and unconnected quarters should thus leap to the same point, can only arise from *that* being the point where truth resides.

Accordingly the cases in which inductions from classes of facts altogether different have thus *jumped together,* belong only to the best established theories which the history of science contains. And as I shall have occasion to refer to this peculiar feature in their evidence, I will take the liberty of describing it by a particular phrase; and will term it the *Consilience of Inductions.*

It is exemplified principally in some of the greatest discoveries. Thus it was found by Newton that the doctrine of the Attraction of the Sun varying according to the Inverse Square of this distance, which explained Kepler's *Third Law* of the proportionality of the cubes of the distances to the squares of the periodic times of the planets, explained also his *First* and *Second Laws* of the elliptical motion of each planet; although no connexion of these laws had been visible before. Again, it appeared that the force of Universal Gravitation, which had been inferred from the *Perturbations* of the moon and planets by the sun and by each other, also accounted for the fact, apparently altogether dissimilar and remote, of the *Precession of the equinoxes.* Here was a most striking and surprizing coincidence, which gave to the theory a stamp of truth beyond the power of ingenuity to counterfeit. In like manner in Optics; the hypothesis of alternate Fits of easy Transmission and Reflection would explain the colours of thin plates, and indeed was devised and adjusted for that very purpose; but it could give no account of the phenomena of the fringes of shadows. But the doctrine of Interferences, constructed at first with reference to phenomena of the nature of the *Fringes,* explained also the *Colours of thin plates* better than the supposition of the fits invented for that very purpose. And we have in Physical Optics another example of the same kind, which is quite as striking as the explanation of precession by inferences from the facts of perturbation. The doctrine of Undulations propagated in a Spheroidal Form was contrived at first by Huyghens, with a view to explain the laws of *Double Refraction* in calc-spar; and was pursued with the same view by Fresnel. But in the course of the investigation it appeared, in a most unexpected and wonderful manner, that this same doctrine of spheroidal undulations, when it was so modified as to account for the directions of the two refracted rays, accounted also for the positions of their *Planes of Polarization*[18]; a phenomenon which, taken by itself, it had perplexed previous mathematicians, even to represent.

The Theory of Universal Gravitation, and of the Undulatory Theory of Light, are, indeed, full of examples of this Consilience of Inductions. With regard to the latter, it has been justly asserted by Herschel, that the history of the undulatory theory was a succession of *felicities.*[19] And it is precisely the unexpected coincidences of results drawn from distant parts of the subject which are properly thus described. Thus the Laws of the *Modification of polarization* to which Fresnel was led by his general views, accounted for the Rule respecting the *Angle at which light is polarized,* discovered by Sir D. Brewster.[20] The conceptions of the theory pointed out peculiar *Modifications* of the phenomena when *Newton's rings* were produced by polarized light, which modifications were ascertained to take place in fact, by Arago and Airy.[21] When the beautiful phenomena of *Dipolarized light* were discovered by Arago and Biot, Young was able to declare that they were reducible to the general laws of *Interference* which he had already established.[22] And what was no less striking a confirmation of the truth of the theory, *Measures* of the same element deduced from various classes of facts were found to coincide. Thus the *Length* of a luminiferous undulation, calculated by Young from the measurement of *Fringes* of shadows, was found to agree very nearly with the previous calculation from the colours of *Thin plates*.[23]

No example can be pointed out, in the whole history of science, so far as I am aware, in which this Consilience of Inductions has given testimony in favour of an hypothesis afterwards discovered to be false. If we take one class of facts only, knowing the law which they follow, we may construct an hypothesis, or perhaps several, which may represent them: and as new circumstances are discovered, we may often adjust the hypothesis so as to correspond to these also. But when the hypothesis, of itself and without adjustment for the purpose, gives us the rule and reason of a class of facts not contemplated in its construction, we have a criterion of its reality, which has never yet been produced in favour of falsehood.

12. In the preceding Article I have spoken of the hypothesis with which we compare our facts as being framed *all at once,* each of its parts being included in the original scheme. In reality, however, it often happens that the various suppositions which our system contains are *added* upon occasion of different researches. Thus in the Ptolemaic doctrine of the heavens, new epicycles and eccentrics were added as new inequalities of the motions of the heavenly bodies were discovered; and in the Newtonian doctrine of material rays of light, the supposition that these rays had "fits," was added to explain the colours of thin plates; and the supposition that they had "sides" was introduced on occasion of the phenomena of polarization. In like manner other theories have been built up of parts devised at different times.

This being the mode in which theories are often framed, we have to notice a distinction which is found to prevail in the progress of true and of false theories. In the former class all the additional suppositions *tend to simplicity* and harmony; the new suppositions resolve themselves into the old ones, or at least require only some easy modification of the hypothesis first assumed: the system becomes more coherent as it is further extended. The elements which we require for explaining a new class of facts are already contained in our system. Different members of the theory run together, and we have thus a constant convergence to unity. In false theories, the contrary is the case. The new suppositions are something altogether additional; – not suggested by the original scheme; perhaps difficult to reconcile with it. Every such addition adds to the complexity of the hypothetical system, which at last becomes unmanageable, and is compelled to surrender its place to some simpler explanation.

Such a false theory, for example, was the ancient doctrine of eccentrics and epicycles. It explained the general succession of the Places of the Sun, Moon, and Planets; it would not have explained the proportion of their Magnitudes at different times, if these could have been accurately observed; but this the ancient astronomers were unable to do. When, however, Tycho and other astronomers came to be able to observe the planets accurately in all positions, it was found that *no* combination of *equable* circular motions would exactly represent all the observations. We may see, in Kepler's works, the many new modifications of the epicyclical hypothesis which offered themselves to him; some of which would have agreed with the phenomena with a certain degree of accuracy, but not so great a degree as Kepler, fortunately for the progress of science, insisted upon obtaining. After these epicycles had been thus accumulated, they all disappeared and gave way to the simpler conception of an *elliptical* motion. In like manner, the discovery of new inequalities in the Moon's motions encumbered her system more and more with new machinery, which was at last rejected all at once in favour of the *elliptical* theory. Astronomers could not but suppose themselves in a wrong path, when the prospect grew darker and more entangled at every step.

Again; the Cartesian system of Vortices might be said to explain the primary phenomena of the revolutions of planets about the sun, and satellites about planets. But the elliptical form of the orbits required new suppositions. Bernoulli ascribed this curve to the shape of the planet, operating on the stream of the vortex in a manner similar to the rudder of a boat. But then the motions of the aphelia, and of the nodes, – the perturbations, – even the action of gravity towards the earth, – could not be accounted for without new and independent suppositions; Here was none of the simplicity of truth. The theory of Gravitation, on the other hand, became more simple as the facts to be explained became more numerous. The attraction of the sun accounted for the motions of the planets; the attraction of the planets was the cause of the motion of the satellites. But this being assumed, the perturbations, the motions of the nodes and aphelia, only made it requisite to extend the attraction of the sun to the satellites, and that of the planets to each other: – the tides, the spheroidal form of the earth, the precession, still required nothing more than that the moon and sun should attract the parts of the earth, and that these should attract each other; – so that all the suppositions resolved themselves into the single one, of the universal gravitation of all matter. It is difficult to imagine a more convincing manifestation of simplicity and unity.

Again, to take an example from another science; – the doctrine of Phlogiston brought together many facts in a very plausible manner, – combustion, acidification, and others, – and very naturally prevailed for a while. But the balance came to be used in chemical operations, and the facts of weight as well as of combination were to be accounted for. On the phlogistic theory, it appeared that this could not be done without a new supposition, and *that,* a very strange one; – that phlogiston was an element not only not heavy, but absolutely light, so that it diminished the weight of the compounds into which it entered. Some chemists for a time adopted this extravagant view; but the wiser of them saw, in the necessity of such a supposition to the defence of the theory, an evidence that the hypothesis of an element *phlogiston* was erroneous. And the opposite hypothesis, which taught that oxygen was subtracted, and not phlogiston added, was accepted because it required no such novel and inadmissible assumption.

Again, we find the same evidence of truth in the progress of the Undulatory Theory of light, in the course of its application from one class of facts to another. Thus we explain Reflection and Refraction by undulations; when we come to Thin Plates, the requisite "fits" are already involved in our fundamental hypothesis, for they are the length of an undulation: the phenomena of Diffraction also require such intervals; and the intervals thus required agree exactly with the others in magnitude, so that no new property is needed. Polarization for a moment appears to require some new hypothesis; yet this is hardly the case; for the direction of our vibrations is hitherto arbitrary: – we allow polarization to decide it, and we suppose the undulations to be transverse. Having done this for the sake of Polarization, we turn to the phenomena of Double Refraction, and inquire what new hypothesis they require. But the answer is, that they require none: the supposition of transverse vibrations, which we have made in order to explain Polarization, gives us also the law of Double Refraction. Truth may give rise to such a coincidence; falsehood cannot. Again, the facts of Dipolarization come into view. But they hardly require any new assumption; for the difference of optical elasticity of crystals in different directions, which is already assumed in uniaxal crystals,[24] is extended to biaxal exactly according to the law of symmetry; and this being done, the laws of the phenomena, curious and complex as they are, are fully explained. The phenomena of Circular Polarization by internal reflection, instead of requiring a new hypothesis, are found to be given by an interpretation of an apparently inexplicable result of an old hypothesis. The Circular Polarization of Quartz and its Double Refraction does indeed appear to require a new assumption, but still not one which at all disturbs the form of the theory; and in short, the whole history of this theory is a progress, constant and steady, often striking and startling, from one degree of evidence and consistence to another of higher order.

In the Emission Theory, on the other hand, as in the theory of solid epicycles, we see what we may consider as the natural course of things in the career of a false theory. Such a theory may, to a certain extent, explain the phenomena which it was at first contrived to meet; but every new class of facts requires a new supposition – an addition to the machinery: and as observation goes on, these incoherent appendages accumulate, till they overwhelm and upset the original frame-work. Such has been the hypothesis of the Material Emission of light. In its original form, it explained Reflection and Refraction: but the colours of Thin Plates added to it the Fits of easy Transmission and Reflection; the phenomena of Diffraction further invested the emitted particles with complex laws of Attraction and Repulsion; Polarization gave them Sides: Double Refraction subjected them to peculiar Forces emanating from the axes of the crystal: finally, Dipolarization loaded them with the complex and unconnected contrivance of Moveable Polarization: and even when all this had been done, additional mechanism was wanting. There is here no unexpected success, no happy coincidence, no convergence of principles from remote quarters. The philosopher builds the machine, but its parts do not fit. They hold together only while he presses them. This is not the character of truth.

As another example of the application of the Maxim now under consideration, I may perhaps be allowed to refer to the judgment which, in the History of Thermotics, I have ventured to give respecting Laplace's Theory of Gases. I have stated,[25] that we cannot help forming an unfavourable judgment of this theory, by looking for that great characteristic of true theory; namely, that the hypotheses which were assumed to account for *one class* of facts are found to explain *another class* of a different nature. Thus Laplace's first suppositions explain the connexion of Compression with Density, (the law of Boyle and Mariotte,) and the connexion of Elasticity with Heat, (the law of Dalton and Gay Lussac.) But the theory requires other assumptions when we come to Latent Heat; and yet these new assumptions produce no effect upon the calculations in any application of the theory. When the hypothesis, constructed with reference to the Elasticity and Temperature, is applied to another class of facts, those of Latent Heat, we have no Simplification of the Hypothesis; and therefore no evidence of the truth of the theory.

13. The two last sections of this chapter direct our attention to two circumstances, which tend to prove, in a manner which we may term irresistible, the truth of the theories which they characterize: – the *Consilience of Inductions* from different and separate classes of facts; – and the progressive *Simplification of the Theory* as it is extended to new cases. These two Characters are, in fact, hardly different; they are exemplified by the same cases. For if these Inductions, collected from one class of facts, supply an unexpected explanation of a new class, which is the case first spoken of, there will be no need for new machinery in the hypothesis to apply it to the newly-contemplated facts; and thus, we have a case in which the system does not become more complex when its application is extended to a wider field, which was the character of true theory in its second aspect. The Consiliences of our Inductions give rise to a constant Convergence of our Theory towards Simplicity and Unity.

But, moreover, both these cases of the extension of the theory, without difficulty or new suppositions, to a wider range and to new classes of phenomena, may be conveniently considered in yet another point of view; namely, as successive steps by which we gradually ascend in our speculative views to a higher and higher point of generality. For when the theory, either by the concurrence of two indications, or by an extension without complication, has included a new range of phenomena, we have, in fact, a new induction of a more general kind, to which the inductions formerly obtained are subordinate, as particular cases to a general proposition. We have in such examples, in short, an instance of *successive generalization.* This is a subject of great importance, and deserving of being well illustrated; it will come under our notice in the next chapter.

Notes

1 *Dynamique.*
2 *Outlines,* &c., p. 26.
3 This would be the case, if, t being the number of seconds elapsed, and C some constant quantity, the velocity were expressed by this mathematical formula, $C(\frac{99}{100})^t$.
4 Poisson, *Dynamique.* Ed. 2, Art. 113.
5 *Hist. Ind. Sci.,* B. VI. c. ii. sect. 2.
6 *Dialogo,* III. p. 95.
7 *Dialogo,* III. p. 91.
8 *Dialogo,* III. p. 36.

9 I repeat here remarks made at the end of the *Mechanical Euclid,* p. 178.
10 *Analyt. Prior.,* Lib. II. c. 23. Περὶ τῆς ἐπαγωγῆς.
11 The syllogism here alluded to would be this: –
Mercury, Venus, Mars, describe ellipses about the Sun;
All Planets do what Mercury, Venus, Mars, do;
Therefore all Planets describe ellipses about the Sun.
12 Εἰ οὖν ἀντιστρέφει τὸ Γ τῷ Β καὶ μὴ ὑπερτείνει τὸ μέσον. – ARISTOT. *Ibid.*
13 I here take the liberty of characterizing inventive minds in general in the same phraseology which, in the History of Science, I have employed in reference to particular examples. These expressions are what I have used in speaking of the discoveries of Copernicus – *Hist. Ind. Sci.,* B. V. c. ii.
14 These observations are made on occasion of Kepler's speculations, and are illustrated by reference to his discoveries. – *Hist. Ind. Sci.,* B. V. c. iv. sect. 1.
15 *Hist. Ind. Sci.,* B. V. c. iv. sect. 1.
16 *Hist. Ind. Sci.,* B. VII. c. ii. sect. 3.
17 Deluc, *Modifications de l'Atmosphere,* Partie I.
18 *Hist. Ind. Sci.,* B. IX. c. xi. sect. 4.
19 See *Hist. Ind. Sci.,* B. IX. c. xii.
20 *Ib.,* c. xi. sect. 4.
21 *Ib.,* c. xiii. sect. 6.
22 *Ib.,* c. xi. sect. 5.
23 *Ib.,* c. xi. sect. 2.
24 *Hist. Ind. Sci.,* B. IX. c. xi. sect. 5.
25 *Hist. Ind. Sci.,* B. X. c. iv.

Part 4

JOHN STUART MILL'S *SYSTEM OF LOGIC*

Editorial Headnote

10. J. S. Mill, *A System of Logic* (1843)

John Stuart Mill (1806–1873), now perhaps best known as the co-founder, with Jeremy Bentham (1748–1832), of the utilitarian system of ethics, was also deeply concerned with the nature of human reasoning, especially in the contexts of science and mathematics. Incredibly sweeping in scope, his six-volume *System of Logic* endeavored to present a unified way in which to understand knowledge ranging from the philosophy of language (names, propositions, and natural classifications), to logic (both syllogistic and inductive), and on to mathematics and natural sciences (including laws of nature, experiment, and causation), and even philosophy (such as logical fallacies, the philosophy of mind, and ethics).

While Mill himself shied away from the term, we would now describe his philosophical orientation as radically empiricist.[1] All knowledge – even, as we will see shortly, that in mathematics – is the result either of definitional convention, or of generalizations on the basis of observation. Departing from Herschel and especially from Whewell, there are no necessary truths to be found in any branch of human inquiry.

We must be limited here to Mill's two most significant impacts on future methodological thought. First, and by far most important, are the methods of causal analysis that are now commonly called "Mill's Methods" (though, as we saw, they were to some extent anticipated by Herschel). Mill offers us a variety of ways to answer the following question: How, in practice, ought we to distinguish which causes are responsible for which effects?

In total, Mill identifies five such methods. (1) The Method of Agreement, in which we look to see which antecedents invariably precede a phenomenon; if there is only one common to all the instances of the effect, it must be the cause; (2) The Method of Difference, in which we find (or create) two trials that differ *only* in the presence or absence of one antecedent; if they then differ in the presence or absence of the effect, that antecedent is the cause; (3) The Indirect Method of Difference, in which we use the Method of Agreement to eliminate irrelevant factors, followed by the Method of Difference between (only) the relevant factors that remain to isolate the real cause; (4) The Method of Residues, on which we eliminate the influence of every known acting cause and see which causes remain that could produce the effect at issue; and (5) The Method of Concomitant Variations, where we look for correlations between variations in causes and variations in effects in an attempt to understand causes (such as gravitation or heat) whose action is never absent (and hence for which the other methods don't apply).

Using these methods correctly, to be sure, is difficult, a fact which Mill readily acknowledges. Being certain, for instance, that two cases for the Method of Difference really *only* differ with respect to our proposed cause usually requires carefully controlled experimentation. Using the Method of Concomitant Variations requires that we have a clear and well-founded understanding of the quantitative measure of the phenomena in question, which is not always a given. The five are also not, strictly speaking, entirely independent methods: (3) is a kind of simultaneous application of the Methods of Agreement and Difference (and thus Mill sometimes speaks of having only derived four methods), and (4) is a version of the Method of Difference.

The second of Mill's contributions that we will explore here is his development of an empiricist grounding for mathematical truths. His position is a radical one. The "deductive" or "exact" sciences, including mathematics, are no more strongly supported than the inductive ones; the "character of necessity, ascribed to the truths of mathematics . . . is an illusion" (p. 177). This is so because the objects of mathematical study have a kind of contradictory existence. We know that they are not real things; there is nothing in the world corresponding to a point with no size, or to a line with no thickness. According to everything we know about physics, we know that they're even *impossible* things; no such objects could ever be constructed. We also, according to Mill, make a mistake when we think that we can work with these objects in our minds, as a kind of abstraction: our real thoughts about points and lines, he argues, will be found just to be "copies of the points, lines, circles, and squares" that we know from our everyday life (pp. 178). What mathematics must truly be about, then, is generalizations about the behavior of objects in our everyday experience – that is, it must be an inductive branch of knowledge just like any other science.

This leads Mill to a sophisticated understanding of the nature of formal reasoning in fields like geometry. Formal sciences do indeed follow necessarily from their axioms, only on the presupposition that their axioms are true; the axioms of geometry themselves are not (as would be confirmed some decades later with the development of non-Euclidean geometry) themselves necessary. Whewell, he writes, is mistaken when he argues that features of human reasoning can confer this *a priori* character upon those axioms: no matter how many times we observe that three objects and two objects makes five objects, we can never, by observation, observe that three plus two *must* equal five. The contrary might seem impossible to us, but only because our habit to think that arithmetic is true is so deeply ingrained. The same goes, Mill argues, for Whewell's attempts to render things like Newton's laws of mechanics *a priori* truths; these will forever remain results derived from our experience. A final objection, particularly poignant for the case of arithmetic and algebra, concerns our ability to derive novel truths from mathematical reasoning. For Mill, deductive reasoning, if it amounts only to the transformation of language (from "if A then B," and "A" to "B," e.g.), can never generate new knowledge, in straightforward contradiction with mathematical practice, which does seem to teach us new things about both the mathematical realm and the natural world. Again, the numbers in any mathematical operation, however abstract, "must be numbers of something" in order to be meaningful (p. 187); arithmetic is just as inductive as the rest of human knowledge.

Note

1 For a longer, general introduction to Mill's *System of Logic*, see R. F. McRae, 'Introduction,' in J. S. Mill, *A System of Logic, Books I–III* (Toronto: University of Toronto Press, 1974), pp. xxi–xlviii.

10

A SYSTEM OF LOGIC

John Stuart Mill

Source: John Stuart Mill, *A System of Logic* (1843), "Of Observation and Experiment" and "Of the Four Methods of Experimental Enquiry," pp. 1:437–1:479 and from "Of Demonstration, and Necessary Truths" and "The Same Subject Continued," pp. 1:296–300, 1:311–323, 1:328–330.

CHAPTER VII.

OF OBSERVATION AND EXPERIMENT.

§ 1. IT results from the preceding exposition, that the process of ascertaining what consequents, in nature, are invariably connected with what antecedents, or in other words what phenomena are related to each other as causes and effects, is in some sort a process of analysis. That every fact which begins to exist has a cause, and that this cause must be found somewhere among the facts which immediately preceded its occurrence, may be taken for certain. The whole of the present facts are the infallible result of all past facts, and more immediately of all the facts which existed at the moment previous. Here, then, is a great sequence, which we know to be uniform. If the whole prior state of the entire universe could again recur, it would again be followed by the whole present state. The question is, how to resolve this complex uniformity into the simpler uniformities which compose it, and assign to each portion of the vast antecedent that portion of the consequent which is attendant upon it.

This operation, which we have called analytical, inasmuch as it is the resolution of a complex whole into the component elements, is more than a merely mental analysis. No mere contemplation of the phenomena, and partition of them by the intellect alone, will of itself accomplish the end we have now in view. Nevertheless, such a mental partition is an indispensable first step. The order of nature, as perceived at a first glance, presents at every instant a chaos followed by another chaos. We must decompose each chaos into single facts. We must learn to see in the chaotic antecedent a multitude of distinct antecedents, in the chaotic consequent a multitude of distinct consequents. This, supposing it done, will not of itself tell us on which of the antecedents each consequent is invariably attendant. To determine that point, we must endeavour to effect a separation of the facts from one another, not in our minds only, but in nature. The mental analysis, however, must take place first. And every one knows that in the mode of performing it, one intellect differs immensely from another. It is the essence of the act of observing;

DOI: 10.4324/9781003263814-14

for the observer is not he who merely sees the thing which is before his eyes, but he who sees what parts that thing is composed of. To do this well is a rare talent. One person, from inattention, or attending only in the wrong place, overlooks half of what he sees; another sets down much more than he sees, confounding it with what he imagines, or with what he infers; another takes note of the *kind* of all the circumstances, but being inexpert in estimating their degree, leaves the quantity of each vague and uncertain; another sees indeed the whole, but makes such an awkward division of it into parts, throwing things into one mass which require to be separated, and separating others which might more conveniently be considered as one, that the result is much the same, sometimes even worse, than if no analysis had been attempted at all. It would be possible to point out what qualities of mind, and modes of mental culture, fit a person for being a good observer; that, however, is a question not of Logic, but of the theory of Education, in the most enlarged sense of the term. There is not properly an Art of Observing. There may be rules for observing. But these, like rules for inventing, are properly instructions for the preparation of one's own mind; for putting it into the state in which it will be most fitted to observe, or most likely to invent. They are, therefore, essentially rules of self-education, which is a different thing from Logic. They do not teach how to do the thing, but how to make ourselves capable of doing it. They are an art of strengthening the limbs, not an art of using them.

The extent and minuteness of observation which may be requisite, and the degree of decomposition to which it may be necessary to carry the mental analysis, depend upon the particular purpose in view. To ascertain the state of the whole universe at any particular moment is impossible, but would also be useless. In making chemical experiments, we should not think it necessary to note the position of the planets; because experience has shown, as a very superficial experience is sufficient to show, that in such cases that circumstance is not material to the result: and, accordingly, in the age when men believed in the occult influences of the heavenly bodies, it might have been unphilosophical to omit ascertaining the precise condition of those bodies at the moment of the experiment. As to the degree of minuteness of the mental sub-division; if we were obliged to break down what we observe into its very simplest elements, that is, literally into single facts, it would be difficult to say where we should find them: we can hardly ever affirm that our divisions of any kind have reached the ultimate unit. But this, too, is fortunately unnecessary. The only object of the mental separation is to suggest the requisite physical separation, so that we may either accomplish it ourselves, or seek for it in nature; and we have done enough when we have carried the sub-division as far as the point at which we are able to see what observations or experiments we require. It is only essential, at whatever point our mental decomposition of facts may for the present have stopped, that we should hold ourselves ready and able to carry it farther as occasion requires, and should not allow the freedom of our discriminating faculty to be imprisoned by the swathes and bands of ordinary classification; as was the case with all early speculative inquirers, not excepting the Greeks, to whom it hardly ever occurred that what was called by one abstract name, might, in reality, be several phenomena, or that there was a possibility of decomposing the facts of the universe into any elements but those which ordinary language already recognised.

§ 2. The different antecedents and consequents being, then, supposed to be, so far as the case requires, ascertained and discriminated from one another; we are to inquire which is connected with which. In every instance which comes under our observation, there are many antecedents and many consequents. If those antecedents could not be severed from one another except in thought, or if those consequents never were found apart, it would be impossible for us to distinguish (*à posteriori* at least) the real laws, or to assign to any cause its effect, or to any effect its cause. To do so, we must be able to meet with some of the antecedents apart from the rest, and observe what follows from them; or some of the consequents, and observe by what they are preceded. We must, in short, follow the Baconian rule of *varying the circumstances.* This is, indeed, only the first rule of physical inquiry, and not, as some have thought, the sole rule; but it is the foundation of all the rest.

For the purpose of varying the circumstances, we may have recourse (according to a distinction commonly made) either to observation or to experiment; we may either *find* an instance in nature, suited to our purposes, or, by an artificial arrangement of circumstances, *make* one. The value of the instance depends upon what it is in itself, not upon the mode in which it is obtained: its employment for the purposes of induction depends upon the same principles in the one case and in the other; as the uses of money are the same whether it is inherited or acquired. There is, in short, no difference in kind, no real logical distinction, between the two processes of investigation. There are, however, practical distinctions to which it is of considerable importance to advert.

§ 3. The first and most obvious distinction between Observation and Experiment is, that the latter is an immense extension of the former. It not only enables us to produce a much greater number of variations in the circumstances than nature spontaneously offers, but, moreover, in thousands of cases, to produce the precise *sort* of variation which we are in want of for discovering the law of the phenomenon; a service which nature, being constructed on a quite different scheme from that of facilitating our studies, is seldom so friendly as to bestow upon us. For example, in order to ascertain what principle in the atmosphere enables it to sustain life, the variation we require is that a living animal should be immersed in each component element of the atmosphere separately. But nature does not supply either oxygen or azote in a separate state. We are indebted to artificial experiment for our knowledge that it is the former, and not the latter, which supports respiration; and even for our knowledge of the very existence of the two ingredients.

Thus far the advantage of experimentation over simple observation is universally recognised: all are aware that it enables us to obtain innumerable combinations of circumstances which are not to be found in nature, and so add to nature's experiments a multitude of experiments of our own. But there is another superiority, (or, as Bacon would have expressed it, another prerogative) of instances artificially obtained over spontaneous instances, – of our own experiments over even the same experiments when made by nature, – which is not of less importance, and which is far from being felt and acknowledged in the same degree.

When we can produce a phenomenon artificially, we can take it, as it were, home with us, and observe it in the midst of circumstances with which in all other respects we are accurately acquainted. If we desire to know what are the effects of the cause A, and are able to produce A by any means at our disposal, we can generally determine at our own discretion, so far as is compatible with the nature of the phenomenon A, the whole of the circumstances which shall be present along with it: and thus, knowing exactly the simultaneous state of everything else which is within the reach of A's influence, we have only to observe what alteration is made in that state by the presence of A.

For example, by the electrical machine we can produce in the midst of known circumstances, the phenomena which nature exhibits on a grander scale under the form of lightning and thunder. Now let any one consider what amount of knowledge of the effects and laws of electric agency mankind could ever have obtained from the mere observation of thunder-storms, and compare it with that which they have gained, and may expect to gain, from electrical and galvanic experiments. This example is the more striking, now that we have reason to believe that electric action is of all natural phenomena (except heat) the most pervading and universal, which, therefore, it might antecedently have been supposed could stand least in need of artificial means of production to enable it to be studied; while the fact is so much the contrary, that without the electric machine, the voltaic battery, and the Leyden jar, we should never have suspected the existence of electricity as one of the great agents in nature; the few electric phenomena we should have known of would have continued to be regarded either as supernatural, or as a sort of anomalies and eccentricities in the order of the universe.

When we have succeeded in insulating the phenomenon which is the subject of inquiry, by placing it among known circumstances, we may produce further variations of circumstances to any extent, and of such kinds as we think best calculated to bring the laws of the phenomenon into a clear light. By introducing one well defined circumstance after another into the experiment, we obtain assurance of the manner in which the phenomenon behaves under an indefinite variety of possible circumstances. Thus, chemists, after having obtained some newly-discovered substance in a pure state, (that is, having made sure that there is nothing present which can interfere with and modify its agency,) introduce various other substances, one by one, to ascertain whether it will combine with them, or decompose them, and with what result; and also apply heat, or electricity, or pressure, to discover what will happen to the substance under each of these circumstances.

But if, on the other hand, it is out of our power to produce the phenomenon, and we have to seek for instances in which nature produces it, the task before us is one of quite another kind. Instead of being able to choose what the concomitant circumstances shall be, we now have to discover what they are; which, when we go beyond the simplest and most accessible cases, it is next to impossible to do, with any precision and completeness. Let us take, as an exemplification of a phenomenon which we have no means of fabricating artificially, a human mind. Nature produces many; but the consequence of our not being able to produce it by art is, that in every instance in which we see a human mind developing itself, or acting upon other things, we see it surrounded and obscured by an indefinite multitude of unascertainable circumstances, rendering the use of the common experimental methods almost delusive. We may conceive to what extent this is true, if we consider, among other things, that whenever nature produces a human mind, she produces, in close connexion with it, also a body; that is, a vast complication of physical facts, in no two cases perhaps exactly similar, and most of which (except the mere structure, which we can examine in a sort of coarse way after it has ceased to act) are radically out of the reach of our means of exploration. If, instead of a human mind, we suppose the subject of investigation to be a human society or State, all the same difficulties recur in a greatly augmented degree.

We have thus already come within sight of a conclusion, which the progress of the inquiry will, I think, bring before us with the clearest evidence: namely, that in the sciences which deal with phenomena in which artificial experiments are impossible (as in the case of astronomy), or in which they have a very limited range (as in physiology, mental philosophy, and the social science,) induction from direct experience is practised at a disadvantage generally equivalent to impracticability: from which it follows that the methods of those sciences, in order to accomplish anything worthy of attainment, must be to a great extent, if not principally, deductive. This is already known to be the case with the first of the sciences we have mentioned, astronomy; that it is not generally recognised as true of the others, is probably one of the reasons why they are still in their infancy. But any further notice of this topic would at present be premature.

§ 4. If what is called pure Observation is at so great a disadvantage compared with artificial experimentation, in one department of the direct exploration of phenomena, there is another branch in which the advantage is all on the side of the former.

Inductive inquiry having for its object to ascertain what causes are connected with what effects, we may begin this search at either end of the road which leads from the one point to the other: we may either inquire into the effects of a given cause, or into the causes of a given effect. The fact that light blackens chloride of silver might have been discovered, either by experiments upon light, trying what effect it would produce on various substances, or by observing that portions of the chloride had repeatedly become black, and inquiring into the circumstances. The effect of the urali poison might have become known either by administering it to animals, or by examining how it happened that the wounds which the Indians of Guiana inflict with their arrows prove so uniformly mortal. Now it is manifest from the mere statement of the examples, without any theoretical discussion, that artificial experimentation is applicable only to the former of these modes of investigation. We can take a cause, and try what it will produce: but we cannot take an effect, and try what it will be produced by. We can only watch till we see it produced, or are enabled to produce it by accident.

This would be of little importance, if it always depended upon our choice from which of the two ends of the sequence we would undertake our inquiries. But we have seldom any option. As we can only travel from the known to the unknown, we are obliged to commence at whichever end we are best acquainted with. If the agent is more familiar to us than its effects, we watch for, or contrive, instances of the agent, under such varieties of circumstances as are open to us, and observe the result. If, on the contrary, the conditions on which a phenomenon depends are obscure, but the phenomenon itself familiar, we must commence our inquiry from the effect. If we are struck with the fact that chloride of silver has been blackened, and have no suspicion of the cause, we have no resource but to compare instances in which the fact has chanced to occur, until by that comparison we discover that in all those instances the substance had been exposed to the light. If we knew nothing of the Indian arrows but their fatal effect, accident alone could turn our attention to experiments on the urali: in the regular course of investigation, we could only inquire, or try to observe, what had been done to the arrows in particular instances.

Wherever, having nothing to guide us to the cause, we are obliged to set out from the effect, and to apply the rule of varying the circumstances to the consequents, not the antecedents, we are necessarily destitute of the resource of artificial experimentation. We cannot, at our choice, obtain consequents, as we can antecedents, under any set of circumstances compatible with their nature. There are no means of producing effects but through their causes, and by the supposition the causes of the effect in question are not known to us. We have therefore no expedient but to study it where it offers itself spontaneously. If nature happens to present us with instances sufficiently varied in their circumstances, and if we are able to discover either among the proximate antecedents, or among some other order of antecedents, something which is always found when the effect is found, however various the circumstances, and never found when it is not; we may discover, by mere observation without experiment, a real uniformity in nature.

But although this is certainly the most favourable case for sciences of pure observation, as contrasted with those in which artificial experiments are possible, there is in reality no case which more strikingly illustrates the inherent imperfection of direct induction when not founded upon experimentation. Suppose that, by a comparison of cases of the effect, we have found an antecedent which appears to be, and perhaps is, invariably connected with it: we have not yet proved that antecedent to be the cause, until we have reversed the process, and produced the effect by means of that antecedent. If we can produce the antecedent artificially, and if, when we do so, the effect follows, the induction is complete; that antecedent is the cause of that consequent.[1] But we then have added the evidence of experiment to that of simple observation. Until we had done so, we had only proved *invariable* antecedence, but not *unconditional* antecedence, or causation. Until it had been shown by the actual production of the antecedent under known circumstances, and the occurrence thereupon of the consequent, that the antecedent was really the condition on which it depended; the uniformity of succession which was proved to exist between them might, for aught we knew, be (like the succession of day and night) no case of causation at all; both antecedent and consequent might be successive stages of the effect of an ulterior cause. Observation, in short, without experiment (and without any aid from deduction) can ascertain uniformities, but cannot prove causation.

In order to see these remarks verified by the actual state of the sciences, we have only to think of the condition of natural history. In zoology, for example, there is an immense number of uniformities ascertained, some of coexistence, others of succession, to many of which, notwithstanding considerable variations of the attendant circumstances, we know not any exception: but the antecedents, for the most part, are such as we cannot artificially produce; or, if we can, it is only by setting in motion the exact process by which nature produces them; and this being to us a mysterious process, of which the main circumstances are not only unknown but unobservable, the name of experimentation would here be completely misapplied. Such are the facts: and what is the result? That on this vast subject, which affords so much and such varied scope for observation, we have not, properly speaking, ascertained a single cause, a single unconditional uniformity. We know not, in the case of almost any of the phenomena that we find conjoined, which is the condition of the other; which is cause, and which effect, or whether either of them is so, or they are not rather all of them conjunct effects of causes yet to be discovered, complex results of laws hitherto unknown.

Although some of the foregoing observations may be, in technical strictness of arrangement, premature in this place, it seemed that a few general remarks upon the difference between Sciences of mere Observation and Sciences of Experimentation, and the extreme disadvantage under which directly inductive inquiry is necessarily carried on in the former, were the best preparation for discussing the methods of direct induction; a preparation rendering superfluous much that must otherwise have been introduced, with some inconvenience, into the heart of that discussion. To the consideration of these Methods we now proceed.

Chapter VIII.

OF THE FOUR METHODS OF EXPERIMENTAL INQUIRY.

§ 1. The simplest and most obvious modes of singling out from among the circumstances which precede or follow a phenomenon, those with which it is really connected by an invariable law, are two in number. One is, by comparing together different instances in which the phenomenon occurs. The other is by comparing instances in which the phenomenon does occur, with instances in other respects similar in which it does not. These two methods may be respectively denominated, the Method of Agreement, and the Method of Difference.

In illustrating these methods it will be necessary to bear in mind the two-fold character of inquiries into the laws of phenomena; which may be either inquiries into the cause of a given effect, or into the effects or properties of a given cause. We shall consider the methods in their application to either order of investigation, and shall draw our examples equally from both.

We shall denote antecedents by the large letters of the alphabet, and the consequents corresponding to them by the small. Let A, then, be an agent or cause, and let the object of our inquiry be to ascertain what are the effects of this cause. If we can either find, or produce, the agent A in such varieties of circumstances, that the different cases have no circumstance in common except A; then, whatever effect we find to be produced in all our trials must, it would seem, be the effect of A. Suppose, for example, that A is tried along with B and C, and that the effect is *a b c;* and suppose that A is next tried with D and E, but without B and C, and that the effect is *a d e*. Then we may reason thus: *b* and *c* are not effects of A, for they were not produced by it in the second experiment; nor are *d* and *e,* for they were not produced in the first. Whatever is really the effect of A must have been produced in both instances; now this condition is fulfilled by no circumstance except *a*. The phenomenon *a* cannot have been the effect of B or C, since it was produced where they were not; nor of D or E, since it was produced where they were not. Therefore it is the effect of A.

For example, let the antecedent A be the contact of an alkaline substance and an oil. This combination being tried under several varieties of circumstance resembling each other in nothing else, the results agree in the production of a greasy and detersive or saponaceous substance: it is therefore concluded that the combination of an oil and an alkali causes the production of a soap. It is thus we inquire, by the Method of Agreement, into the effect of a given cause.

In a similar manner we may inquire into the cause of a given effect. Let *a* be the effect. Here, as shown in the last chapter, we have only the resource of observation without experiment: we cannot take a phenomenon of which we know not the origin, and try to find its mode of production by producing it; if we succeeded in such a random trial it could only be by accident. But if we can observe *a* in two different combinations, *a b c* and *a d e;* and if we know, or can discover, that the antecedent circumstances in these cases respectively were A B C and A D E; we may conclude by a reasoning similar to that in the preceding example, that A is the antecedent connected with the consequent *a* by a law of causation. B and C, we may say, cannot be causes of *a,* since on its second occurrence they were not present; nor are D and E, for they were not present on its first occurrence. A, alone of the five circumstances, was found among the antecedents of *a* in both instances.

For example, let the effect *a* be crystallization. We compare instances in which bodies are known to assume crystalline structure, but which have no other point of agreement; and we find them to have one, and as far as we can observe, only one, antecedent in common: the deposition of a solid matter from a liquid state, either a state of fusion or of solution. We conclude, therefore, that the solidification of a substance from a liquid state is an invariable antecedent of its crystallization.

In this example we may go farther, and say, it is not only the invariable antecedent but the cause. For in this case we are able, after detecting the antecedent A, to produce it artificially, and by finding that *a* follows it, verify the result of our induction. The importance of thus reversing the proof was never more strikingly manifested than when, by keeping a phial of water charged with siliceous particles undisturbed for years, a chemist (I believe Dr. Wollaston) succeeded in obtaining crystals of quartz; and in the equally interesting experiment in which Sir James Hall produced artificial marble, by the cooling of its materials from fusion under immense pressure: two admirable examples of the light which may be thrown upon the most secret processes of nature by well-contrived interrogation of her.

But if we cannot artificially produce the phenomenon A, the conclusion that it is the cause of *a* remains subject to very considerable doubt. Though an invariable, it may not be the unconditional antecedent of *a,* but may precede it as day precedes night or night day. This uncertainty arises from the impossibility of assuring ourselves that A is the *only* immediate antecedent common to both the instances. If we could be certain of having ascertained all the invariable antecedents, we might be sure that the unconditional invariable antecedent, or cause, must be found somewhere among them. Unfortunately it is hardly ever possible to ascertain all the antecedents, unless the phenomenon is one which we can produce artificially. Even then the difficulty is merely lightened, not removed: men knew how to raise water in pumps long before they adverted to what was really the operating circumstance in the means they employed, namely, the pressure of the atmosphere on the open surface of the water. It is, however, much easier to analyze completely a set of arrangements made by ourselves, than the whole complex mass of the agencies which nature happens to be exerting at the moment when she produces any given phenomenon. We may overlook some of the material circumstances in an experiment with an electrical machine; but we shall, at the worst, be better acquainted with them than with those of a thunder-storm.

The mode of discovering and proving laws of nature, which we have now examined, proceeds upon the following axiom: Whatever circumstance can be excluded, without prejudice to the phenomenon, or can be absent notwithstanding its presence, is not connected with it in the way of causation. The casual circumstances being thus eliminated, if only one remains, that one is the cause which we are in search of: if more than one, they either are, or contain among them, the cause: and so, *mutatis mutandis,* of the effect. As this method proceeds by comparing different instances to ascertain in what they agree, I have termed it the Method of Agreement: and we may adopt as its regulating principle the following canon: –

First Canon.

If two or more instances of the phenomenon under investigation have only one circumstance in common, the circumstance in which alone all the instances agree, is the cause (or effect) of the given phenomenon.

Quitting for the present the Method of Agreement, to which we shall almost immediately return, we proceed to a still more potent instrument of the investigation of nature, the Method of Difference.

§ 2. In the Method of Agreement, we endeavoured to obtain instances which agreed in the given circumstance but differed in every other: in the present method we require, on the contrary, two instances resembling one another in every other respect, but differing in the presence or absence of the phenomenon we wish to study. If our object be to discover the effects of an agent A, we must procure A in some set of ascertained circumstances, as A B C, and having noted the effects produced, compare them with the effect of the remaining circumstances B C, when A is absent. If the effect of A B C is *a b c,* and the effect of B C, *b c,* it is evident that the effect of A is *a.* So again, if we begin at the other end, and desire to investigate the cause of an effect *a,* we must select an instance, as *a b c,* in which the effect occurs, and in which the antecedents were A B C, and we must look put for another instance in which the remaining circumstances, *b c,* occur without *a.* If the antecedents, in that instance, are B C, we know that the cause of *a* must be A: either A alone, or A in conjunction with some of the other circumstances present.

It is scarcely necessary to give examples of a logical process to which we owe almost all the inductive conclusions we draw in daily life. When a man is shot through the heart, it is by this method we know that it was the gun-shot which killed him: for he was in the fulness of life immediately before, all circumstances being the same, except the wound.

The axioms which are taken for granted in this method are evidently the following: Whatever antecedent cannot be excluded without preventing the phenomenon, is the cause, or a condition, of that phenomenon; Whatever consequent can be excluded, with no other difference in the antecedents than the absence of a particular one, is the effect of that one. Instead of comparing different instances of a phenomenon, to discover in what they agree, this method compares an instance of its occurrence with an instance of its non-occurrence, to discover in what they differ. The canon which is the regulating principle of the Method of Difference may be expressed as follows: –

Second Canon.

> *If an instance in which the phenomenon under investigation occurs, and an instance in which it does not occur, have every circumstance save one in common, that one occurring only in the former; the circumstance in which alone the two instances differ, is the effect, or cause, or a necessary part of the cause, of the phenomenon.*

§ 3. The two methods which we have now stated have many features of resemblance, but there are also many distinctions between them. Both are methods of *elimination.* This term (which is employed in the theory of equations to denote the process by which one after another of the elements of a question is excluded, and the solution made to depend upon the relation between the remaining elements only,) is well suited to express the operation, analogous to this, which has been understood since the time of Bacon to be the foundation of experimental inquiry: namely, the successive exclusion of the various circumstances which are found to accompany a phenomenon in a given instance, in order to ascertain what are those among them which can be absent consistently with the existence of the phenomenon. The Method of Agreement stands on the ground that whatever can be eliminated, is not connected with the phenomenon by any law. The Method of Difference has for its foundation, that whatever can *not* be eliminated, *is* connected with the phenomenon by a law.

Of these methods, that of Difference is more particularly a method of artificial experiment; while that of Agreement is more especially the resource we employ where experimentation is impossible. A few reflections will prove the fact, and point out the reason of it.

It is inherent in the peculiar character of the Method of Difference, that the nature of the combinations which it requires is much more strictly defined than in the Method of Agreement. The two instances which are to be compared with one another must be exactly similar, in all circumstances except the one which we are attempting to investigate: they must be in the relation of A B C and B C, or of *a b c* and *b c*. It is true that this similarity of circumstances needs not extend to such as are already known to be immaterial to the result. And in the case of most phenomena we learn at once, from the most ordinary experience, that most of the coexistent phenomena of the universe may be either present or absent without affecting the given phenomenon; or, if present, are present indifferently when the phenomenon does not happen, and when it does. Still, even limiting the identity which is required between the two instances, A B C and B C, to such circumstances as are not already known to be indifferent; it is very seldom that nature affords two instances, of which we can be assured that they stand in this precise relation to one another. In the spontaneous operations of nature there is generally such complication and such obscurity, they are mostly either on so overwhelmingly large or on so inaccessibly minute a scale, we are so ignorant of a great part of the facts which really take place, and even those of which we are not ignorant are so multitudinous, and therefore so seldom exactly alike in any two cases, that a spontaneous experiment, of the kind required by the Method of Difference, is commonly not to be found. When, on the contrary, we obtain a phenomenon by an artificial experiment, a pair of instances such as the method requires is obtained almost as a matter of course, provided the process does not last a long time. A certain state of surrounding circumstances existed before we commenced the experiment: this is B C. We then introduce A; say, for instance, by merely bringing an object from another part of the room, before there has been time for any change in the other elements. It is, in short (as M. Comte observes), the very nature of an experiment, to introduce into the preexisting state of circumstances a change perfectly definite. We choose a previous state of things with which we are well acquainted, so that no unforeseen alteration in that state is likely to pass unobserved; and into this we introduce, as rapidly as possible, the phenomenon which we wish to study; so that we in general are entitled to feel complete assurance, that the pre-existing state, and the state which we have produced, differ in nothing except in the presence or absence of that phenomenon. If a bird is taken from a cage, and instantly plunged into carbonic acid gas, the experimentalist may be fully assured (at all events after one or two repetitions) that no circumstance capable of causing suffocation had supervened in the interim, except the change from immersion in the atmosphere to immersion in carbonic acid gas. There is one doubt, indeed, which may remain in some cases of this description; the effect may have been produced not by the change, but by the means we employed to produce the change. The possibility, however, of this last supposition generally admits of being conclusively tested by other experiments. It thus appears that in the study of the various kinds of phenomena which we can, by our voluntary agency, modify or control, we can in general satisfy the requisitions of the Method of Difference; but that by the spontaneous operations of nature those requisitions are seldom fulfilled.

The reverse of this is the case with the Method of Agreement. We do not here require instances of so special and determinate a kind. Any instances whatever, in which nature presents us with a phenomenon, may be examined for the purposes of this method; and if *all* such instances agree in anything, a conclusion of considerable value is already attained. We can seldom, indeed, be sure that this one point of agreement is the only one; but our ignorance does not, as in the Method of Difference, vitiate the conclusion; the certainty of the result, as far as it goes, is not affected. We have ascertained one invariable antecedent or consequent, however many other invariable antecedents or consequents may still remain unascertained. If A B C, A D E, A F G, are all equally followed by *a,* then *a* is an invariable consequent of A. If *a b c, a d e, a f g* all number A among their antecedents, then A is connected as an antecedent, by some invariable law, with *a.* But to determine whether this invariable antecedent is a cause, or this invariable consequent an effect, we must be able, in addition, to produce the one by means of the other; or, at least, to obtain that which alone constitutes our assurance of having produced anything, namely, an instance in which the effect, *a,* has come into existence, with no other change in the pre-existing circumstances than the addition of A. And this, if we can do it, is an application of the Method of Difference, not of the Method of Agreement.

It thus appears to be by the Method of Difference alone that we can ever, in the way of direct experience, arrive with certainty at causes. The Method of Agreement leads only to laws of phenomena, as Mr. Whewell calls them, but which (since laws of causation are also laws of phenomena) I prefer to designate as uniformities in which the question of causation must for the present remain undecided. The Method of Agreement is chiefly to be resorted to, as a means of suggesting applications of the Method of Difference (as in the last example the comparison of A B C, A D E, A F G, suggested that A was the antecedent on which to try the experiment whether it could produce *a*); or as an inferior resource, in case the Method of Difference is impracticable; which, as we before showed, generally arises from the impossibility of artificially producing the phenomena.

And hence it is that the Method of Agreement, although applicable in principle to either case, is more emphatically the method of investigation on those subjects where artificial experimentation is impossible; because on those it is, generally, our only resource of a directly inductive nature; while, in the phenomena which we can produce at pleasure, the Method of Difference generally affords a more efficacious process, which will ascertain causes as well as mere laws.

§ 4. Our next remark shall be, that there are many cases in which, although our power of producing the phenomenon is complete, the Method of Difference either cannot be made available at all, or not without a previous employment of the Method of Agreement. This occurs when the agency by which we can produce the phenomenon is not that of one single antecedent, but a combination of antecedents, which we have no power of separating from each other and exhibiting apart. For instance, suppose the subject of inquiry to be the cause of the double refraction of light. We can produce this phenomenon at pleasure, by employing any one of the many substances which are known to refract light in that peculiar manner. But if, taking one of those substances, as Iceland spar for example, we wish to determine on which of the properties of Iceland spar this remarkable phenomenon depends, we can make no use, for that purpose, of the Method of Difference; for we cannot find another substance precisely resembling Iceland spar except in some one property. The only mode, therefore, of prosecuting this inquiry is that afforded by the Method of Agreement; by which, in fact, through a comparison of all the known substances which had the property of doubly refracting light, it was ascertained that they agreed in the single circumstance of being crystalline substances; and although the converse does not hold, although all crystalline substances have not the property of double refraction, it was concluded, with reason, that there is a real connection between these two properties; that either crystalline structure, or the cause which gives rise to that structure, is one of the conditions of double refraction.

Out of this employment of the Method of Agreement arises a peculiar modification of that method, which is sometimes of great avail in the investigation of nature. In cases similar to the above, in which it is not possible to obtain the precise pair of instances which our second canon requires – instances agreeing in every antecedent except A, or in every consequent except *a*; we may yet be able, by a double employment of the Method of Agreement, to discover in what the instances which contain A or *a,* differ from those which do not.

If we compare various instances in which *a* occurs, and find that they all have in common the circumstance A, and (as far as can be observed) no other circumstance, the Method of Agreement, so far, bears testimony to a connexion between A and *a*. In order to convert this proof of connexion into proof of causation by the direct Method of Difference, we ought to be able in some one of these instances, as for example A B C, to leave out A, and observe whether by doing so, *a* is prevented. Now supposing (what is often the case) that we are not able to try this decisive experiment; yet, provided we can by any means discover what would be its result if we *could* try it, the advantage will be the same. Suppose, then, that as we previously examined a variety of instances in which *a* occurred, and found them to agree in containing A, so we now observe a variety of instances in which *a* does *not* occur, and find them agree in *not* containing A; which establishes, by the Method of Agreement, the same connexion between the absence of A and the absence of *a,* which was before established between their presence. As, then, it had been shown that whenever A is present *a* is present, so it being now shown that when A is taken away *a* is removed along with it, we have by the one proposition A B C, *a b c*, by the other B C, *b c,* the positive and negative instances which the Method of Difference requires. Thus, if it be true that all animals which have a well-developed respiratory system, and therefore aerate the blood perfectly, agree in being warm-blooded, while those whose respiratory system is imperfect do not maintain a temperature much exceeding that of the surrounding medium, we may argue from this twofold experience, that the change which takes place in the blood by respiration is the cause of animal heat.

This method may be called the Indirect Method of Difference, or the Joint Method of Agreement and Difference; and consists in a double employment of the Method of Agreement, each proof being independent of the other, and corroborating it. But it is not equivalent to a proof by the direct Method of Difference. For the requisitions of the Method of Difference are not satisfied, unless we can be quite sure either that the instances affirmative of *a* agree in no antecedent whatever but A, or that the instances negative of *a* agree in nothing but the negation of A. Now if it were possible, which it never is, to have this assurance, we should not need the joint method; for either of the two sets of instances separately would then be sufficient to prove causation. This indirect method, therefore, can only be viewed as a great extension and improvement of the Method of Agreement, but not as participating in the more cogent nature of the Method of Difference. The following may be stated as its canon: –

Third Canon.

If two or more instances in which the phenomenon occurs have only one circumstance in common, while two or more instances in which it does not occur have nothing in common save the absence of that circumstance; the circumstance in which alone the two sets of instances differ, is the effect, or cause, or a necessary part of the cause, of the phenomenon.

We shall presently show that the Joint Method of Agreement and Difference constitutes, in another respect not yet adverted to, an improvement upon the common Method of Agreement, namely, in being unaffected by a characteristic imperfection of that method, the nature of which still remains to be pointed out. But as we cannot enter into this exposition without introducing a new element of complexity into this long and intricate discussion, I shall postpone it to the next chapter, and shall at once proceed to the statement of two other methods, which will complete the enumeration of the means which mankind possess for exploring the laws of nature by specific observation and experience.

§ 5. The first of these has been aptly denominated the Method of Residues. Its principle is very simple. Subducting from any given phenomenon all the portions which, by virtue of preceding inductions, can be assigned to known causes, the remainder will be the effect of the antecedents which had been overlooked, or of which the effect was as yet an unknown quantity.

Suppose, as before, that we have the antecedents A B C, followed by the consequents *a b c,* and that by previous inductions, (founded, we will suppose, upon the Method of Difference,) we have ascertained the causes of some of these effects, or the effects of some of these causes; and are by this means apprised that the effect of A is *a,* and that the effect of B is *b.* Subtracting the sum of these effects from the total phenomenon, there remains *c,* which now, without any fresh experiment, we may know to be the effect of C. This Method of Residues is in truth a peculiar modification of the Method of Difference. If the instance A B C*, a b c*, could have been compared with a single instance A B*, a b*, we should have proved C to be the cause of *c*, by the common process of the Method of Difference. In the present case, however, instead of a single instance A B, we have had to study separately the causes A and B, and to infer from the effects which they produce separately, what effect they must produce in the case A B C where they act together. Of the two instances, therefore, which the Method of Difference requires, – the one positive, the other negative, – the negative one, or that in which the given phenomenon is absent, is not the direct result of observation and experiment, but has been arrived at by deduction. As one of the forms of the Method of Difference, the Method of Residues partakes of its rigorous certainty, provided the previous inductions, those which gave the effects of A and B, were obtained by the same infallible method, and provided we are certain that C is the antecedent to which the residual phenomenon *c* can be referred; the only agent of which we had not already calculated and subducted the effect. But as we can never be quite certain of this, the evidence derived from the Method of Residues is not complete, unless we can obtain C artificially and try it separately, or unless its agency, when once suggested, can be accounted for, and proved deductively, from known laws.

Even with these reservations, the Method of Residues is one of the most important among our instruments of discovery. Of all the methods of investigating laws of nature, this is the most fertile in unexpected results; often informing us of sequences in which neither the cause nor the effect were sufficiently conspicuous to attract of themselves the attention of observers. The agent C may be an obscure circumstance, not likely to have been perceived unless sought for, nor likely to have been sought for until attention had been awakened by the insufficiency of the obvious causes to account for the whole of the effect. And *c* may be so disguised by its intermixture with *a* and *b,* that it would scarcely have presented itself spontaneously as a subject of separate study. Of these uses of the method, we shall presently cite some remarkable examples. The canon of the Method of Residues is as follows: –

Fourth Canon.

Subduct from any phenomenon such part as is known by previous inductions to be the effect of certain antecedents, and the residue of the phenomenon is the effect of the remaining antecedents.

§ 6. There remains a class of laws which it is impracticable to ascertain by any of the three methods which I have attempted to characterize; namely, the laws of those Permanent Causes, or indestructible natural agents, which it is impossible either to exclude or to isolate; which we can neither hinder from being present, nor contrive that they shall be present alone. It would appear at first sight that we could by no means separate the effects of these agents from the effects of those other phenomena with which they cannot be prevented from co-existing. In respect, indeed, to most of the permanent causes, no such difficulty exists; since, though we cannot eliminate them as coexisting facts, we can eliminate them as influencing agents, by simply trying our experiment in a local situation beyond the limits of their influence. The pendulum, for example, has its oscillations disturbed by the vicinity of a mountain: we remove the pendulum to a sufficient distance from the mountain, and the disturbance ceases: from these data we can determine by the Method of Difference, the amount of effect really due to the mountain; and beyond a certain distance everything goes on precisely as it would do if the mountain exercised no influence whatever, which, accordingly, we, with sufficient reason, conclude to be the fact.

The difficulty, therefore, in applying the methods already treated of to determine the effects of Permanent Causes, is confined to the cases in which it is impossible for us to get out of the local limits of their influence. The pendulum can be removed from the influence of the mountain, but it cannot be removed from the influence of the earth: we cannot take away the earth from the pendulum, nor the pendulum from the earth, to ascertain whether it would continue to vibrate if the action which the earth exerts upon it were withdrawn. On what evidence, then, do we ascribe its vibrations to the earth's influence? Not on any sanctioned by the Method of Difference; for one of the two instances, the negative instance, is wanting. Nor by the Method of Agreement; for although all pendulums agree in this, that during their oscillations the earth is always present, why may we not as well ascribe the phenomenon to the sun, which is equally a coexistent fact in all the experiments? It is evident that to establish even so simple a fact of causation as this, there was required some method over and above those which we have yet examined.

As another example, let us take the phenomenon Heat. Independently of all hypothesis as to the real nature of the agency so called, this fact is certain, that we are unable to exhaust any body of the whole of its heat. It is equally certain that no one ever perceived heat not emanating from a body. Being unable, then, to separate Body and Heat, we cannot effect such a variation of circumstances as the foregoing three methods require; we cannot ascertain, by those methods, what portion of the phenomena exhibited by any body are due to the heat contained in it. If we could observe a body with its heat, and the same body entirely divested of heat, the Method of Difference would show the effect due to the heat, apart from that due to the body. If we could observe heat under circumstances agreeing in nothing but heat, and therefore not characterized also by the presence of a body, we could ascertain the effects of heat, from an instance of heat with a body and an instance of heat without a body, by the Method of Agreement; or, if we pleased, we could determine by the Method of Difference what effect was due to the body, when the remainder which was due to the heat would be given by the Method of Residues. But we can do none of these things; and without them the application of any of the three methods to the solution of this problem would be illusory. It would be idle, for instance, to attempt to ascertain the effect of heat by subtracting from the phenomena exhibited by a body, all that is due to its other properties; for as we have never been able to observe any bodies without a portion of heat in them, the effects due to that heat may form a part of the very results, which we affect to subtract in order that the effect of heat may be shown by the residue.

If, therefore, there were no other methods of experimental investigation than these three, we should be for ever unable to determine the effects due to heat as a cause. But we have still a resource. Though we cannot exclude an antecedent altogether, we may be able to produce, or nature may produce for us, some modification in it. By a modification is here meant, a change in it, not amounting to its total removal. If some modification in the antecedent A is always followed by a change in the consequent *a,* the other consequents *b* and *c* remaining the same; or, *vice versâ,* if every change in *a* is found to have been preceded by some modification in A, none being observable in any of the other antecedents; we may safely conclude that *a* is, wholly or in part, an effect traceable to A, or at least in some way connected with it through causation. For example, in the case of heat, though we cannot expel it altogether from any body, we can modify it in quantity, we can increase or diminish it; and doing so, we find by the various methods of experimentation or observation already treated of, that such increase or diminution of heat is followed by expansion or contraction of the body. In this manner we arrive at thc conclusion, otherwise unattainable by us, that one of the effects of heat is to enlarge the dimensions of bodies; or what is the same thing in other words, to widen the distances between their particles.

A change in a thing, not amounting to its total removal, that is, a change which leaves it still the same thing it was, must be a change either in its quantity, or in some of its relations to other things, of which relations the principal is its position in space. In the previous example, the modification which was produced in the antecedent was an alteration in its quantity. Let us now suppose the question to be, what influence the moon exerts on the surface of the earth. We cannot try an experiment in the absence of the moon, so as to observe what terrestrial phenomena her annihilation would put an end to; but when we find that all the variations in the *position* of the moon are followed by corresponding variations in the time and place of high water, the place being always either on the side of the earth which is nearest to, or on that which is most remote from, the moon, we have ample evidence that the moon is, wholly or partially, the cause which determines the tides. It very commonly happens, as it does in this instance, that the variations of an effect are correspondent, or analogous, to those of its cause; as the moon moves further towards the east, the high water point does the same: but this is not an indispensable condition; as may be seen in the same example, for along with that high water point, there is at the same instant another high water point diametrically opposite to it, and which, therefore, of necessity, moves towards the west as the moon followed by the nearer of the tide waves advances towards the east: and yet both these motions are equally effects of the moon's motion.

That the oscillations of the pendulum are caused by the earth, is proved by similar evidence. Those oscillations take place between equidistant points on the two sides of a line, which, being perpendicular to the earth, varies with every variation in the earth's position, either in space or relatively to the object. Speaking accurately, we only know by the method now characterized, that all terrestrial bodies tend to the earth, and not to some unknown fixed point lying in the same direction. In every twenty-four hours, by the earth's rotation, the line drawn from the body at right angles to the earth coincides successively with all the radii of a circle, and in the course of six months the place of that circle varies by nearly two hundred millions of miles; yet in all these changes of the earth's position, the line in which bodies tend to fall continues to be directed towards it: which proves that terrestrial gravity is directed to the earth, and not, as was once fancied by some, to a fixed point of space.

The method by which these results were obtained, may be termed the Method of Concomitant Variations: it is regulated by the following canon: –

Fifth Canon.

Whatever phenomenon varies in any manner whenever another phenomenon varies in some particular manner, is either a cause or an effect of that phenomenon, or is connected with it through some fact of causation.

The last clause is subjoined, because it by no means follows when two phenomena accompany each other in their variations, that the one is cause and the other effect. The same thing may, and indeed must happen, supposing them to be two different effects of a common cause: and by this method alone it would never be possible to ascertain which of the two suppositions is the true one. The only way to solve the doubt would be that which we have so often adverted to, viz., by endeavouring to ascertain whether we can produce the one set of variations by means of the other. In the case of heat, for example, by increasing the temperature of a body we increase its bulk, but by increasing its bulk we do not increase its temperature; on the contrary, (as in the rarefaction of air under the receiver of an air-pump,) we generally diminish it: therefore heat is not an effect, but a cause, of increase of bulk. If we cannot ourselves produce the variations, we must endeavour, though it is an attempt which is seldom successful, to find them produced by nature in some case in which the pre-existing circumstances are perfectly known to us.

It is scarcely necessary to say, that in order to ascertain the uniform concomitance of variations in the effect with variations in the cause, the same precautions must be used as in any other case of the determination of an invariable sequence. We must endeavour to retain all the other antecedents unchanged, while that particular one is subjected to the requisite series of variations; or in other words, that we may be warranted in inferring causation from concomitance of variations, the concomitance itself must be proved by the Method of Difference.

It might at first appear that the Method of Concomitant Variations assumes a new axiom, or law of causation in general, namely, that every modification of the cause is followed by a change in the effect. And it does usually happen that when a phenomenon A causes a phenomenon *a,* any variation in the quantity or in the various relations of A, is uniformly followed by a variation in the quantity or relations of *a*. To take a familiar instance, that of gravitation. The sun causes a certain tendency to motion in the earth; here we have cause and effect; but that tendency is *towards* the sun, and therefore varies in direction as the sun varies in the relation of position; and moreover the tendency varies in intensity, in a certain numerical ratio to the sun's distance from the earth, that is, according to another relation of the sun. Thus we see that there is not only an invariable connexion between the sun and the earth's gravitation, but that two of the relations of the sun, its position with respect to the earth and its distance from the earth, are invariably connected as antecedents with the quantity and direction of the earth's gravitation. The cause of the earth's gravitating at all, is simply the sun; but the cause of her gravitating with a given intensity and in a given direction, is the existence of the sun in a given direction and at a given distance. It is not strange that a modified cause, which is in truth a different cause, should produce a different effect. But as the cause is only different in its quantity, or in some of its relations, it usually happens that the effect also is only changed in its quantity or its relations.

Although it is for the most part true that a modification of the cause is followed by a modification of the effect, the Method of Concomitant Variations does not, however, pre-suppose this as an axiom. It only requires the converse proposition; that anything upon whose modifications, modifications of an effect are invariably consequent, must be the cause (or connected with the cause) of that effect; a proposition, the truth of which is evident; for if the thing itself had no influence on the effect, neither could the modications of the thing have any influence. If the stars have no power over the fortunes of men, it is implied in the very terms, that the conjunctions or oppositions of different stars can have no such power.

Although the most striking applications of the Method of Concomitant Variations take place in the cases in which the Method of Difference, strictly so called, is impossible, its use is not confined to those cases; it may often usefully follow *after* the Method of Difference, to give additional precision to a solution which that has found. When by the Method of Difference it has first been ascertained that a certain object produces a certain effect, the Method of Concomitant Variations may be usefully called in to determine according to what law the quantity or the different relations of the effect follow those of the cause.

§ 7. The case in which this method admits of the most extensive employment, is that in which the variations of the cause are variations of quantity. Of such variations we may in general affirm with safety, that they will be attended not only with variations, but with similar variations, of the effect: the proposition, that more of the cause is followed by more of the effect, being a corollary from the principle of the Composition of Causes, which, as we have seen, is the general rule of causation; cases of the opposite description, in which causes change their properties on being conjoined with one another, being, on the contrary, special and exceptional. Suppose, then, that when A changes in quantity, *a* also changes in quantity, and in such a manner that we can trace the numerical relation which the changes of the one bear to such changes of the other as take place within our limits of observation. We may then, with certain precautions, safely conclude that the same numerical relation will hold beyond those limits. If, for instance, we find that when A is double, *a* is double; that when A is treble or quadruple, *a* is treble or quadruple; we may conclude that if A were a half or a third, *a* would be a half or a third, and finally, that if A were annihilated, *a* would be annihilated, and that *a* is wholly the effect of A, or wholly the effect of the same cause with A. And so with any other numerical relation according to which A and *a* would vanish simultaneously; as for instance if *a* were proportional to the square of A. If, on the other hand, *a* is not wholly the effect of A, but yet varies when A varies, it is probably (to use a mathematical phrase) a function not of A alone but of A and something else: its changes will be such as would occur if part of it remained constant, or varied on some other principle, and the remainder varied in some numerical relation to the variations of A. In that case, when A diminishes, *a* will seem to approach not towards zero, but towards some other limit: and when the series of variations is such as to indicate what that limit is, if constant, or the law of its variation if variable, the limit will exactly measure how much of *a* is the effect of some other and independent cause, and the remainder will be the effect of A (or of the cause of A).

These conclusions, however, must not be drawn without certain precautions. In the first place, the possibility of drawing them at all, manifestly supposes that we are acquainted not only with the variations, but with the absolute quantities, both of A and *a*. If we do not know the total quantities, we cannot, of course, determine the real numerical relation according to which those quantities vary. It is therefore an error to conclude, as some have concluded, that because increase of heat expands bodies, that is, increases the distance between their particles, therefore that distance is wholly the effect of heat, and that if we could entirely exhaust the body of its heat, the particles would be in complete contact. This can never be more than a guess, and of the most hazardous sort, not a legitimate induction: for since we neither know how much heat there is in any body, nor what is the real distance between any two of its particles, we cannot judge whether the contraction of the distance does or does not follow the diminution of the quantity of heat according to such a numerical relation that the two quantities would vanish simultaneously.

In contrast with this, let us consider a case in which the absolute quantities are known; the case contemplated in the first law of motion; viz., that all bodies in motion continue to move in a straight line with uniform velocity until acted upon by some new force. This assertion is in open opposition to first appearances; all terrestrial objects, when in motion, gradually abate their velocity and at last stop; which accordingly the ancients, with their *inductio per enumerationem simplicem,* imagined to be the law. Every moving body, however, encounters various obstacles, as friction, the resistance of the atmosphere, &c., which we know by daily experience to be causes capable of destroying motion. It was suggested that the whole of the retardation might be owing to these causes. How was this inquired into? If the obstacles could have been entirely removed, the case would have been amenable to the Method of Difference. They could not be removed, they could only be diminished, and the case, therefore, admitted only of the Method of Concomitant Variations. This accordingly being employed, it was found that every diminution of the obstacles diminished the retardation of the motion: and inasmuch as in this case (unlike the case of heat) the total quantities both of the antecedent and of the consequent were known; it was practicable to estimate, with an approach to accuracy, both the amount of the retardation and the amount of the retarding causes, or resistances, and to judge how near they both were to being exhausted; and it appeared that the effect dwindled as rapidly, and at each step was as far on the road towards annihilation, as the cause was. The simple oscillation of a weight suspended from a fixed point, and moved a little out of the perpendicular, which in ordinary circumstances lasts but a few minutes, was prolonged in Borda's experiments to more than thirty hours, by diminishing as much as possible the friction at the point of suspension, and by making the body oscillate in a space exhausted as nearly as possible of its air. There could therefore be no hesitation in assigning the whole of the retardation of motion to the influence of the obstacles: and since, after subducting this retardation from the total phenomenon, the remainder was an uniform velocity, the result was the proposition known as the first law of motion.

There is also another characteristic uncertainty affecting the inference that the law of variation which the quantities observe within our limits of observation, will hold beyond those limits. There is of course, in the first instance, the possibility that beyond the limits, and in circumstances therefore of which we have no direct experience, some counteracting cause might develop itself; either a new agent, or a new property of the agents concerned, which lies dormant in the circumstances we are able to observe. This is an element of uncertainty which enters largely into all our predictions of effects; but it is not peculiarly applicable to the Method of Concomitant Variations. The uncertainty, however, of which I am about to speak, is characteristic of that method; especially in the cases in which the extreme limits of our observation are very narrow, in comparison with the possible variations in the quantities of the phenomena. Any one who has the slightest acquaintance with mathematics, is aware that very different laws of variation may produce numerical results which differ but slightly from one another within narrow limits; and it is often only when the absolute amounts of variation are considerable, that the difference between the results given by one law and by another, becomes appreciable. When, therefore, such variations in the quantity of the antecedents as we have the means of observing, are but small in comparison with the total quantities, there is much danger lest we should mistake the numerical law, and be led quite to miscalculate the variations which would take place beyond the limits; a miscalculation which would vitiate any conclusion respecting the dependence of the effect upon the cause, which could be founded upon those variations. Examples are not wanting of such mistakes. "The formulæ," says Sir John Herschel,[2] "which have been empirically deduced for the elasticity of steam (till very recently), and those for the resistance of fluids, and other similar subjects," when relied on beyond the limits of the observations from which they were deduced, "have almost invariably failed to support the theoretical structures which have been erected on them."

Under this uncertainty, the conclusion we may draw from the concomitant variations of *a* and A, to the existence of an invariable and exclusive connexion between them, or to the permanency of the same numerical relation between their variations when the quantities are much greater or smaller than those which we have had the means of observing, cannot be considered to rest upon a complete induction. All that in such a case can be regarded as proved on the subject of causation, is that there is some connexion between the two phenomena; that A, or something which can influence A, must be of the causes which collectively determine *a*. We may, however, feel assured that the relation which we have observed to exist between the variations of A and *a*, will hold true in all cases which fall between the same extreme limits; that is, wherever the utmost increase or diminution in which the result has been found by observation to coincide with the law, is not exceeded.

The four methods which it has now been attempted to describe, are the only possible modes of experimental inquiry, of direct induction *à posteriori,* as distinguished from deduction: at least I know not, nor am able to conceive, any others. And even of these, the Method of Residues, as we have seen, is not independent of deduction; though, as it requires specific experience in addition, it may, without impropriety, be included among methods of direct observation and experiment.

These, then, with such assistance as can be obtained from Deduction, compose the available resources of the human mind for ascertaining the laws of the succession of phenomena. Before proceeding to point out certain circumstances, by which the employment of these methods is subjected to an immense increase of complication and of difficulty, it is expedient to illustrate the use of the methods, by suitable examples, drawn from actual physical investigations. These, accordingly, will form the subject of the succeeding chapter.

[...]

Chapter V.

OF DEMONSTRATION, AND NECESSARY TRUTHS.

§ 1. If, as laid down in the two preceding chapters, the foundation of all sciences, even deductive or demonstrative sciences, is Induction; if every step in the ratiocinations even of geometry is an act of induction; and if a train of reasoning is but bringing many inductions to bear upon the same subject of inquiry, and drawing a case within one induction by means of another; wherein lies the peculiar certainty always ascribed to the sciences which are entirely, or almost entirely, deductive? Why are they called the Exact Sciences? Why are mathematical certainty, and the evidence of demonstration, common phrases to express the very highest degree of assurance attainable by reason? Why are mathematics by almost all philosophers, and (by many) even those branches of natural philosophy which, through the medium of mathematics, have been converted into deductive sciences, considered to be independent of the evidence of experience and observation, and characterised as systems of Necessary Truth?

The answer I conceive to be, that this character of necessity, ascribed to the truths of mathematics, and even (with some reservations to be hereafter made) the peculiar certainty attributed to them, is an illusion; in order to sustain which, it is necessary to suppose that those truths relate to, and express the properties of, purely imaginary objects. It is acknowledged that the conclusions of geometry are deduced, partly at least, from the so-called Definitions, and that those definitions are assumed to be correct descriptions, as far as they go, of the objects with which geometry is conversant. Now we have pointed out that, from a definition as such, no proposition, unless it be one concerning the meaning of a word, can ever follow; and that what apparently follows from a definition, follows in reality from an implied assumption that there exists a real thing conformable thereto. This assumption, in the case of the definitions of geometry, is false: there exist no real things exactly conformable to the definitions. There exist no points without magnitude; no lines without breadth, nor perfectly straight; no circles with all their radii exactly equal, nor squares with all their angles perfectly right. It will perhaps be said that the assumption does not extend to the actual, but only to the possible, existence of such things. I answer that, according to any test we have of possibility, they are not even possible. Their existence, so far as we can form any judgment, would seem to be inconsistent with the physical constitution of our planet at least, if not of the universe. To get rid of this difficulty, and at the same time to save the credit of the supposed systems of necessary truth, it is customary to say that the points, lines, circles, and squares which are the subject of geometry, exist in our conceptions merely, and are part of our minds; which minds, by working on their own materials, construct an *à priori* science, the evidence of which is purely mental, and has nothing whatever to do with outward experience. By howsoever high authorities this doctrine may have been sanctioned, it appears to me psychologically incorrect. The points, lines, circles, and squares, which any one has in his mind, are (I apprehend) simply copies of the points, lines, circles, and squares which he has known in his experience. A line as defined by geometers is wholly inconceivable. We can reason about a line as if it had no breadth; because we have a power, which is the foundation of all the control we can exercise over the operations of our minds; the power, when a perception is present to our senses, or a conception to our intellects, of *attending* to a part only of that perception or conception, instead of the whole. But we cannot *conceive* a line without breadth; we can form no mental picture of such a line: all the lines which we have in our minds are lines possessing breadth. If any one doubts this, we may refer him to his own experience. I much question if any one who fancies that he can conceive what is called a mathematical line, thinks so from the evidence of his consciousness: I suspect it is rather because he supposes that unless such a conception were possible, mathematics could not exist as a science: a supposition which there will be no difficulty in showing to be entirely groundless.

Since then neither in nature, nor in the human mind, do there exist any objects exactly corresponding to the definitions of geometry, while yet that science cannot be supposed to be conversant about non-entities; nothing remains but to consider geometry as conversant with such lines, angles, and figures as really exist; and the definitions, as they are called, must be regarded as some of our first and most obvious generalizations concerning those natural objects. The correctness of those generalizations, *as* generalizations, is without a flaw: the equality of all the radii of a circle is true of all circles, so far as it is true of any one: but it is not exactly true of any circle: it is only nearly true; so nearly that no error of any importance in practice will be incurred by feigning it to be exactly true. When we have occasion to extend these inductions, or their consequences, to cases in which the error would be appreciable – to lines of perceptible breadth or thickness, parallels which deviate sensibly from equidistance, and the like – we correct our conclusions, by combining with them a fresh set of propositions relating to the aberration; just as we also take in propositions relating to the physical or chemical properties of the material, if those properties happen to introduce any modification into the result, which they easily may, even with respect to figure and magnitude, as in the case, for instance, of expansion by heat. So long, however, as there exists no practical necessity for attending to any of the properties of the object except its geometrical properties, or to any of the natural irregularities in those, it is convenient to neglect the consideration of the other properties and of the irregularities, and to reason as if these did not exist: accordingly, we formally announce, in the definitions, that we intend to proceed on this plan. But it is an error to suppose, because we resolve to confine our attention to a certain number of the properties of an object, that we therefore conceive, or have an idea of, the object, denuded of its other properties. We are thinking, all the time, of precisely such objects as we have seen and touched, and with all the properties which naturally belong to them; but, for scientific convenience, we feign them to be divested of all properties, except those in regard to which we design to consider them.

The peculiar accuracy, supposed to be characteristic of the first principles of geometry, thus appears to be fictitious. The assertions on which the reasonings of the science are founded, do not, any more than in other sciences, exactly correspond with the fact; but we *suppose* that they do so, for the sake of tracing the consequences which follow from the supposition. The opinion of Dugald Stewart respecting the foundations of geometry, is, I conceive, substantially correct; that it is built upon hypotheses; that it owes to this alone the peculiar certainty supposed to distinguish it; and that in any science whatever, by reasoning from a set of hypotheses, we may obtain a body of conclusions as certain as those of geometry, that is, as strictly in accordance with the hypotheses, and as irresistibly compelling assent *on condition* that those hypotheses are true.

When, therefore, it is affirmed that the conclusions of geometry are necessary truths, the necessity consists in reality only in this, that they necessarily follow from the suppositions from which they are deduced. Those suppositions are so far from being necessary, that they are not even true; they purposely depart, more or less widely, from the truth. The only sense in which necessity can be ascribed to the conclusions of any scientific investigation, is that of necessarily following from some assumption, which, by the conditions of the inquiry, is not to be questioned. In this relation, of course, the derivative truths of every deductive science must stand to the inductions, or assumptions, on which the science is founded, and which, whether true or untrue, certain or doubtful in themselves, are always supposed certain for the purposes of the particular science. And therefore the conclusions of all deductive sciences were said by the ancients to be necessary propositions. We have observed already that to be predicated necessarily was characteristic of the predicable Proprium, and that a proprium was any property of a thing which could be deduced from its essence, that is, from the properties included in its definition.

[. . .]

§ 6. The first of the two great arguments in support of the theory that axioms are *à priori* truths, having, I think, been sufficiently answered; I proceed to the second, on which most stress is usually laid, and which is chiefly insisted upon by Mr. Whewell. Axioms (it is asserted) are conceived by us not only as true, but as universally and necessarily true. Now, experience cannot possibly give to any proposition this character. I may have seen snow a hundred times, and may have seen that it was white, but this cannot give me entire assurance even that all snow is white much less that snow *must* be white. "However many instances we may have observed of the truth of a proposition, there is nothing to assure us that the next case shall not be an exception to the rule. If it be strictly true that every ruminant animal yet known has cloven hoofs, we still cannot be sure that some creature will not hereafter be discovered which has the first of these attributes, without having the other. . . . Experience must always consist of a limited number of observations; and, however numerous these may be, they can show nothing with regard to the infinite number of cases in which the experiment has not been made." Moreover, axioms are not only universal, they are also necessary. Now "experience cannot offer the smallest ground for the necessity of a proposition. She can observe and record what has happened; but she cannot find, in any case, or in any accumulation of cases, any reason for what *must* happen. She may see objects side by side; but she cannot see a reason why they must ever be side by side. She finds certain events to occur in succession; but the succession supplies, in its occurrence, no reason for its recurrence. She contemplates external objects; but she cannot detect any internal bond, which indissolubly connects the future with the past, the possible with the real. To learn a proposition by experience, and to see it to be necessarily true, are two altogether different processes of thought."[3] And Mr. Whewell adds, "If any one does not clearly comprehend this distinction of necessary and contingent truths, he will not be able to go along with us in our researches into the foundations of human knowledge; nor, indeed, to pursue with success any speculation on the subject."[4]

In order to learn what the distinction is, the nonrecognition of which incurs this denunciation, let us refer again to Mr. Whewell. "Necessary truths are those in which we not only learn that the proposition *is* true, but see that it *must be* true; in which the negation of the truth is not only false, but impossible; in which we cannot, even by an effort of imagination, or in a supposition, conceive the reverse of that which is asserted. That there are such truths cannot be doubted. We may take, for example, all relations of number. Three and Two, added together, make Five. We cannot conceive it to be otherwise. We cannot, by any freak of thought, imagine Three and Two to make Seven."[5]

Although Mr. Whewell has naturally and properly employed a variety of phrases to bring his meaning more forcibly home, he will, I presume, allow that they are all equivalent; and that what he means by a necessary truth, would be sufficiently defined, a proposition the negation of which is not only false but inconceivable. I am unable to find in any of Mr. Whewell's expressions, turn them what way you will, a meaning beyond this, and I do not believe he would contend that they mean anything more.

This, therefore, is the principle asserted: that propositions, the negation of which is inconceivable, or in other words, which we cannot figure to ourselves as being false, must rest upon evidence of a higher and more cogent description than any which experience can afford. And we have next to consider whether there is any ground for this assertion.

Now I cannot but wonder that so much stress should be laid upon the circumstance of inconceivableness, when there is such ample experience to show, that our capacity or incapacity of conceiving a thing has very little to do with the possibility of the thing in itself; but is in truth very much an affair of accident, and depends upon the past history and habits of our own minds. There is no more generally acknowledged fact in human nature, than the extreme difficulty at first felt in conceiving anything as possible, which is in contradiction to long established and familiar experience; or even to old and familiar habits of thought. And this difficulty is a necessary result of the fundamental laws of the human mind.

When we have often seen and thought of two things together, and have never in any one instance either seen or thought of them separately, there is by the primary law of association an increasing difficulty, which in the end becomes insuperable, of conceiving the two things apart. This is most of all conspicuous in uneducated persons, who are in general utterly unable to separate any two ideas which have once become firmly associated in their minds; and if persons of cultivated intellect have any advantage on the point, it is only because, having seen and heard and read more, and being more accustomed to exercise their imagination, they have experienced their sensations and thoughts in more varied combinations, and have been prevented from forming many of these inseparable associations. But this advantage has necessarily its limits. The man of the most practised intellect is not exempt from the universal laws of our conceptive faculty. If daily habit presents to him for a long period two facts in combination, and if he is not led during that period either by accident or intention to think of them apart, he will in time become incapable of doing so even by the strongest effort; and the supposition that the two facts can be separated in nature, will at last present itself to his mind with all the characters of an inconceivable phenomenon. There are remarkable instances of this in the history of science: instances, in which the wisest men rejected as impossible, because inconceivable, things which their posterity, by earlier practice and longer perseverance in the attempt, found it quite easy to conceive, and which everybody now knows to be true. There was a time when men of the most cultivated intellects, and the most emancipated from the dominion of early prejudice, could not credit the existence of antipodes; were unable to conceive, in opposition to old association, the force of gravity acting upwards instead of downwards. The Cartesians long rejected the Newtonian doctrine of the gravitation of all bodies towards one another, on the faith of a general proposition, the reverse of which seemed to them to be inconceivable – the proposition that a body cannot act where it is not. All the cumbrous machinery of imaginary vortices, assumed without the smallest particle of evidence, appeared to these philosophers a more rational mode of explaining the heavenly motions, than one which involved what seemed to them so great an absurdity.[6] And they no doubt found it as impossible to conceive that a body should act upon the earth, at the distance of the sun or moon, as we find it to conceive an end to space or time, or two straight lines inclosing a space. Newton himself had not been able to realize the conception, or we should not have had his hypothesis of a subtle ether, the occult cause of gravitation; and his writings prove, that although he deemed the particular nature of the intermediate agency a matter of conjecture, the necessity of *some* such agency appeared to him indubitable. It would seem that even now the majority of scientific men have not completely got over this very difficulty; for though they have at last learnt to conceive the sun *attracting* the earth without any intervening fluid, they cannot yet conceive the sun *illuminating* the earth without some such medium.

If, then, it be so natural to the human mind, even in its highest state of culture, to be incapable of conceiving, and on that ground to believe impossible, what is afterwards not only found to be conceivable but proved to be true; what wonder if in cases where the association is still older, more confirmed, and more familiar, and in which nothing ever occurs to shake our conviction, or even suggest to us any conception at variance with the association, the acquired incapacity should continue, and be mistaken for a natural incapacity? It is true our experience of the varieties in nature enables us, within certain limits, to conceive other varieties analogous to them. We can conceive the sun or moon falling; for although we never saw them fall, nor ever perhaps imagined them falling, we have seen so many other things fall, that we have innumerable familiar analogies to assist the conception; which, after all, we should probably have some difficulty in framing, were we not well accustomed to see the sun and moon move (or appear to move), so that we are only called upon to conceive a slight change in the direction of motion, a circumstance familiar to our experience. But when experience affords no model on which to shape the new conception, how is it possible for us to form it? How, for example, can we imagine an end to space or time? We never saw any object without something beyond it, nor experienced any feeling without something following it. When, therefore, we attempt to conceive the last point of space, we have the idea irresistibly raised of other points beyond it. When we try to imagine the last instant of time, we cannot help conceiving another instant after it. Nor is there any necessity to assume, as is done by the school to which Mr. Whewell belongs, a peculiar fundamental law of the mind to account for the feeling of infinity inherent in our conceptions of space and time; that apparent infinity is sufficiently accounted for by simpler and universally acknowledged laws.

Now, in the case of a geometrical axiom, such, for example, as that two straight lines cannot inclose a space, – a truth which is testified to us by our very earliest impressions of the external world, – how is it possible (whether those external impressions be or be not the ground of our belief) that the reverse of the proposition *can* be otherwise than inconceivable to us? What analogy have we, what similar order of facts in any other branch of our experience, to facilitate to us the conception of two straight lines inclosing a space? Nor is even this all. I have already called attention to the peculiar property of our impressions of form, that the ideas or mental images exactly resemble their prototypes, and adequately represent them for the purposes of scientific observation. From this, and from the intuitive character of the observation, which in this case reduces itself to simple inspection, we cannot so much as call up in our imagination two straight lines, in order to attempt to conceive them inclosing a space, without by that very act repeating the philosophical experiment which establishes the contrary. Will it really be contended that the inconceivableness of the thing, under such circumstances, proves anything against the experimental origin of the conviction? Is it not clear that in whichever mode our belief in the proposition may have originated, the impossibility of our conceiving the negative of it must, under either hypothesis, be the same? As, then, Mr. Whewell exhorts those who have any difficulty in recognising the distinction held by him between necessary and contingent truths, to study geometry, – a condition which I can assure him I have conscientiously fulfilled, – I, in return, with equal confidence, exhort those who agree with Mr. Whewell, to study the elementary laws of association; being convinced that nothing more is requisite than a moderate familiarity with those laws, to dispel the illusion which ascribes a peculiar necessity to our earliest inductions from experience, and measures the possibility of things in themselves, by the human capacity of conceiving them.

I hope to be pardoned for adding, that Mr. Whewell himself has both confirmed by his testimony the effect of habitual association in giving to an experimental truth the appearance of a necessary one, and afforded a striking instance of that remarkable law in his own person. In his *Philosophy of the Inductive Sciences* he continually asserts, that propositions which not only are not self-evident, but which we know to have been discovered gradually, and by great efforts of genius and patience, have, when once established, appeared so self-evident that, but for historical evidence, it would have been impossible to conceive that they had not been recognised from the first by all persons in a sound state of their faculties. "We now despise those who, in the Copernican controversy, could not conceive the apparent motion of the sun on the heliocentric hypothesis; or those who, in opposition to Galileo, thought that a uniform force might be that which generated a velocity proportional to the space; or those who held there was something absurd in Newton's doctrine of the different refrangibility of differently coloured rays; or those who imagined that when elements combine, their sensible qualities must be manifest in the compound; or those who were reluctant to give up the distinction of vegetables into herbs, shrubs, and trees. We cannot help thinking that men must have been singularly dull of comprehension to find a difficulty in admitting what is to us so plain and simple. We have a latent persuasion that we in their place should have been wiser and more clearsighted; that we should have taken the right side, and given our assent at once to the truth. Yet in reality such a persuasion is a mere delusion. The persons who, in such instances as the above, were on the losing side, were very far in most cases from being persons more prejudiced, or stupid, or narrow-minded, than the greater part of mankind now are; and the cause for which they fought was far from being a manifestly bad one, till it had been so decided by the result of the war. So complete has been the victory of truth in most of these instances, that at present we can hardly imagine the struggle to have been necessary. *The very essence of these triumphs is, that they lead us to regard the views we reject as not only false, but inconceivable.*[7]"

This last proposition is precisely what I contend for; and I ask no more, in order to overthrow the whole theory of Mr. Whewell on the nature of the evidence of axioms. For what is that theory? That the truth of axioms cannot have been learnt from experience, because their falsity is inconceivable. But Mr. Whewell himself says, that we are continually led by the natural progress of thought, to regard as inconceivable what our forefathers not only conceived but believed, nay, even (he might have added) were unable to conceive the contrary of. Mr. Whewell cannot intend to justify this mode of thought; he cannot mean to say, that we can be *right* in regarding as inconceivable what others have conceived, and as self-evident what to others did not appear evident at all. After so complete an admission that inconceivableness is an accidental thing, not inherent in the phenomenon itself, but dependent on the mental history of the person who tries to conceive it, how can he ever call upon us to reject a proposition as impossible on no other ground than its inconceivableness? Yet he not only does so, but has unintentionally afforded some of the most remarkable examples which can be cited of the very illusion which he has himself so clearly pointed out. We select as specimens, his remarks on the evidence of the three laws of motion, and of the atomic theory.

With respect to the laws of motion, Mr. Whewell says: "No one can doubt that, in historical fact, these laws were collected from experience. That such is the case, is no matter of conjecture. We know the time, the persons, the circumstances, belonging to each step of each discovery."[8] After such a testimony, to adduce evidence of the fact would be superfluous. And not only were these laws by no means intuitively evident, but some of them were originally paradoxes. The first law was especially so. That a body, once in motion, would continue for ever to move in the same direction with undiminished velocity unless acted upon by some new force, was a proposition which mankind found for a long time the greatest difficulty in crediting. It stood opposed to apparent experience of the most familiar kind, which taught that it was the nature of motion to abate gradually, and at last terminate of itself. Yet when once the contrary doctrine was firmly established, mathematicians, as Mr. Whewell observes, speedily began to believe that laws, thus contradictory to first appearances, and which, even after full proof had been obtained, it had required generations to render familiar to the minds of the scientific world, were under "a demonstrable necessity, compelling them to be such as they are and no other;" and Mr. Whewell, though he has "not ventured absolutely to pronounce" that *all* these laws "can be rigorously traced to an absolute necessity in the nature of things,"[9] does actually think in that manner of the law just mentioned; of which he says: "Though the discovery of the first law of motion was made, historically speaking, by means of experiment, we have now attained a point of view in which we see that it might have been certainly known to be true, independently of experience."[10] Can there be a more striking exemplification than is here afforded, of the effect of association which we have described? Philosophers, for generations, have the most extraordinary difficulty in putting certain ideas together; they at last succeed in doing so; and after, a sufficient repetition of the process, they first fancy a natural bond between the ideas, then experience a growing difficulty, which at last, by the continuation of the same progress, becomes an impossibility, of severing them from one another. If such be the progress of an experimental conviction of which the date is of yesterday, and which is in opposition to first appearances, how must it fare with those which are conformable to appearances familiar from the first dawn of intelligence, and of the conclusiveness of which, from the earliest records of human thought, no sceptic has suggested even a momentary doubt?

The other instance which we shall quote is a truly astonishing one, and may be called the *reductio ad absurdum* of the theory of inconceivableness. Speaking of the laws of chemical composition, Mr. Whewell says[11]: "That they could never have been clearly understood, and therefore never firmly established, without laborious and exact experiments, is certain; but yet we may venture to say, that being once known, they possess an evidence beyond that of mere experiment. *For how, in fact, can we conceive combinations, otherwise than as definite in kind and quantity?* If we were to suppose each element ready to combine with any other indifferently, and indifferently in any quantity, we should have a world in which all would be confusion and indefiniteness. There would be no fixed kinds of bodies; salts, and stones, and ores, would approach to and graduate into each other by insensible degrees. Instead of this, we know that the world consists of bodies distinguishable from each other by definite differences, capable of being classified and named, and of having general propositions asserted concerning them. And as *we cannot conceive a world in which this should not be the case,* it would appear that we cannot conceive a state of things in which the laws of the combination of elements should not be of that definite and measured kind which we have above asserted."

That a philosopher of Mr. Whewell's eminence should gravely assert that we cannot concieve a world in which the simple elements would combine in other than definite proportions; that by dint of meditating on a scientific truth, the original discoverer of which is still living, he should have rendered the association in his own mind between the idea of combination and that of constant proportions so familiar and intimate as to be unable to conceive the one fact without the other; is so signal an instance of the law of human nature for which I am contending, that one word more in illustration must be quite superfluous. I shall only, therefore, express my satisfaction that so long as the progress of scientific instruction has not rendered this association as indissoluble in the minds of most people as Mr. Whewell finds it, the majority of mankind will be fairly able to judge, from this example, of the value of the evidence which he deems sufficient to prove that a scientific proposition might be known to be true independently of experience.

CHAPTER VI.

THE SAME SUBJECT CONTINUED.

§ 1. IN the examination which formed the subject of the last chapter, into the nature of the evidence of those deductive sciences which are commonly represented to be systems of necessary truth, we have been led to the following conclusions. The results of those sciences are indeed necessary, in the sense of necessarily following from certain first principles, commonly called axioms and definitions; of being certainly true if those axioms and definitions are so. But their claim to the character of necessity in any sense beyond this, as implying an evidence independent of and superior to observation and experience, must depend upon the previous establishment of such a claim in favour of the definitions and axioms themselves. With regard to axioms, we found that, considered as experimental truths, they rest upon superabundant and obvious evidence. We inquired, whether, since this is the case, it be necessary to suppose any other evidence of those truths than experimental evidence, any other origin for our belief of them than an experimental origin. We decided, that the burden of proof lies with those who maintain the affirmative, and we examined, at considerable length, such arguments as they have produced. The examination having led to the rejection of those arguments, we have thought ourselves warranted in concluding that axioms are but a class, the highest class, of inductions from experience: the simplest and easiest cases of generalization from the facts furnished to us by our senses or by our internal consciousness.

While the axioms of demonstrative sciences thus appeared to be experimental truths, the definitions, as they are incorrectly called, of those sciences, were found by us to be generalizations from experience which are not even, accurately speaking, truths; being propositions in which, while we assert of some kind of object, some property or properties which observation shows to belong to it, we at the same time deny that it possesses any other properties, although in truth other properties do in every individual instance accompany, and in most or even in all instances, modify, the property thus exclusively predicated. The denial, therefore, is a mere fiction, or supposition, made for the purpose of excluding the consideration of those modifying circumstances, when their influence is of too trifling amount to be worth considering, or adjourning it, when important, to a more convenient moment.

From these considerations it would appear that Deductive or Demonstrative Sciences are all, without exception, Inductive Sciences; that their evidence is that of experience, but that they are also, in virtue of the peculiar character of one indispensable portion of the general formulæ according to which their inductions are made, Hypothetical Sciences. Their conclusions are only true upon certain suppositions, which are, or ought to be, approximations to the truth, but are seldom, if ever, exactly true; and to this hypothetical character is to be ascribed the peculiar certainty, which is supposed to be inherent in demonstration.

What we have now asserted, however, cannot be received as universally true of Deductive or Demonstrative Sciences until verified by being applied to the most remarkable of all those sciences, that of Numbers; the theory of the Calculus; Arithmetic and Algebra. It is harder to believe of the doctrines of this science than of any other, either that they are not truths *à priori,* but experimental truths, or that their peculiar certainty is owing to their being not absolute but only conditional truths. This, therefore, is a case which merits examination apart; and the more so, because on this subject we have a double set of doctrines to contend with; that of Mr. Whewell and the *à priori* philosophers on one side; and on the other, a philosophical theory the most opposite to theirs, which was at one time very generally received, and is still far from being altogether exploded among metaphysicians.

§ 2. This theory attempts to solve the difficulty apparently inherent in the case, by representing the propositions of the science of numbers as merely verbal, and its processes as simple transformations of language, substitutions of one expression for another. The proposition, Two and one are equal to three, according to these philosophers, is not a truth, is not the assertion of a really existing fact, but a definition of the word three; a statement that mankind have agreed to use the name three as a sign exactly equivalent to two and one; to call by the former name whatever is called by the other more clumsy phrase. According to this doctrine, the longest process in algebra is but a succession of changes in terminology, by which equivalent expressions are substituted one for another; a series of translations of the same fact, from one into another language; though how, after such a series of translations, the fact itself comes out changed, (as when we demonstrate a new geometrical theorem by algebra,) they have not explained; and it is a difficulty which is fatal to their theory.

It must be acknowledged that there are peculiarities in the processes of arithmetic and algebra which render the above theory very plausible, and have not unnaturally made those sciences the stronghold of Nominalism. The doctrine that we can discover facts, detect the hidden processes of nature, by an artful manipulation of language, is so contrary to common sense, that a person must have made some advances in philosophy to believe it; men fly to so paradoxical a belief to avoid, as they think, some even greater difficulty, which the vulgar do not see. What has led many to believe that reasoning is a mere verbal process, is, that no other theory seemed reconcileable with the nature of the Science of Numbers. For we do not carry any ideas along with us when we use the symbols of arithmetic or of algebra. In a geometrical demonstration we have a mental diagram, if not one upon paper; AB, AC, are present to our imagination as lines, intersecting other lines, forming an angle with one another, and the like; but not so *a* and *b*. These may represent lines or any other magnitudes, but those magnitudes are never thought of; nothing is realized in our imagination but *a* and *b*. The ideas which, on the particular occasion, they happen to represent, are banished from the mind during every intermediate part of the process between the beginning, when the premisses are translated from things into signs, and the end, when the conclusion is translated back from signs into things. Nothing, then, being in the reasoner's mind but the symbols, what can seem more inadmissible than to pretend that the reasoning process has to do with anything more? We seem to have come to one of Bacon's Prerogative Instances; an *experimentum crucis* on the nature of reasoning itself.

Nevertheless it will appear on consideration, that this apparently so decisive instance is no instance at all; that there is in every step of an arithmetical or algebraical calculation a real induction, a real inference of facts from facts; and that what disguises the induction is simply its comprehensive nature, and the consequent extreme generality of the language. All numbers must be numbers of something: there are no such things as numbers in the abstract. *Ten* must mean ten bodies, or ten sounds, or ten beatings of the pulse. But though numbers must be numbers of something, they may be numbers of anything. Propositions, therefore, concerning numbers, have the remarkable peculiarity that they are propositions concerning all things whatever; all objects, all existences of every kind, known to our experience. All things possess quantity; consist of parts which can be numbered; and in that character possess all the properties which are called properties of numbers. That half of four is two must be true whatever the word four represents, whether four men, four miles, or four pounds weight. We need only conceive a thing divided into four equal parts, (and all things may be conceived as so divided,) to be able to predicate of it every property of the number four, that is, every arithmetical proposition in which the number four stands on one side of the equation. Algebra extends the generalization still farther: every number represents that particular number of all things without distinction, but every algebraical symbol does more, it represents all numbers without distinction. As soon as we conceive a thing divided into equal parts, without knowing into what number of parts, we may call it *a* or *x,* and apply to it, without danger of error, every algebraical formula in the books. The proposition, $2\ (a + b) = 2\ a + 2\ b$, is a truth co-extensive with the creation. Since then algebraical truths are true of all things whatever, and not, like those of geometry, true of lines only or angles only, it is no wonder that the symbols should not excite in our minds ideas of any things in particular. When we demonstrate the forty-seventh proposition of Euclid, it is not necessary that the words should raise in us an image of all right-angled triangles, but only of some one right-angled triangle: so in algebra we need not, under the symbol *a,* picture to ourselves all things whatever, but only some one thing; why not, then, the letter itself? The mere written characters, *a, x, y, z,* serve as well for representatives of Things in general, as any more complex and apparently more concrete conception. That we are conscious of them however in their character of things, and not of mere signs, is evident from the fact that our whole process of reasoning is carried on by predicating of them the properties of things. In resolving an algebraic equation, by what rules do we proceed? By applying at each step to *a, b,* and *x,* the proposition that equals added to equals make equals; that equals taken from equals leave equals; and other propositions founded upon these two. These are not properties of language, or of signs as such, but of magnitudes, which is as much as to say, of all things. The inferences, therefore, which are successively drawn, are inferences concerning Things, not symbols; although as any Things whatever will serve the turn, there is no necessity for keeping the idea of the Thing at all distinct, and consequently the process of thought may, in this case, be allowed without danger to do what all processes of thought, when they have been performed often, will do if permitted, namely, to become entirely mechanical. Hence the general language of algebra comes to be used familiarly without exciting ideas, as all other general language is prone to do from mere habit, though in no other case than this can it be done with complete safety. But when we look back to see from whence the probative force of the process is derived, we find that at every single step, unless we suppose ourselves to be thinking and talking of the things, and not the mere symbols, the evidence fails.

There is another circumstance, which, still more than that which we have now mentioned, gives plausibility to the notion that the propositions of arithmetic and algebra are merely verbal. This is, that when considered as propositions respecting Things, they all have the appearance of being identical propositions. The assertion, Two and one are equal to three, considered as an assertion respecting objects, as for instance "Two pebbles and one pebble are equal to three pebbles," does not affirm equality between two collections of pebbles, but absolute identity. It affirms that if we put one pebble to two pebbles, those very pebbles are three. The objects, therefore, being the very same, and the mere assertion that "objects are themselves" being insignificant, it seems but natural to consider the proposition, Two and one are equal to three, as asserting mere identity of signification between the two names.

This, however, though it looks so plausible, will not stand examination. The expression, "two pebbles and one pebble," and the expression, "three pebbles," stand indeed for the same aggregation of objects, but they by no means stand for the same physical fact. They are names of the same objects, but of those objects in two different states: though they denote the same things, their connotation is different. Three pebbles in two separate parcels, and three pebbles in one parcel, do not make the same impression on our senses; and the assertion that the very same pebbles may by an alteration of place and arrangement be made to produce either the one set of sensations or the other, though it is a very familiar proposition, is not an identical one. It is a truth known to us by early and constant experience: an inductive truth; and such truths are the foundation of the science of Number. The fundamental truths of that science all rest upon the evidence of sense; they are proved by showing to our eyes and our fingers that any given number of objects, ten balls for example, may by separation and re-arrangement exhibit to our senses all the different sets of numbers the sum of which is equal to ten. All the improved methods of teaching arithmetic to children proceed upon a knowledge of this fact. All who wish to carry the child's *mind* along with them in learning arithmetic; all who (as Dr. Biber in his remarkable *Lectures on Education* expresses it) wish to teach numbers, and not mere ciphers – now teach it through the evidence of the senses, in the manner we have described.[12]

We may, if we please, call the proposition "Three is two and one," a definition of the number three, and assert that arithmetic, as it has been asserted that geometry, is a science founded upon definitions. But they are definitions in the geometrical sense, not the logical; asserting not the meaning of a term only, but along with it an observed matter of fact. The proposition, "A circle is a figure bounded by a line which has all its points equally distant from a point within it," is called the definition of a circle; but the proposition from which so many consequences follow, and which is really a first principle of geometry, is, that figures answering to this description exist. And thus we may call, "Three is two and one," a definition of three; but the calculations which depend upon that proposition do not follow from the definition itself, but from an arithmetical theorem presupposed in it, namely, that collections of objects exist, which while they impress the senses thus, ∘∘∘, may be separated into two parts, thus, ∘∘ ∘. This proposition being granted, we term all such parcels Threes, after which the enunciation of the above-mentioned physical fact will serve also for a definition of the word Three.

The Science of Number is thus no exception to the conclusion we previously arrived at, that the processes even of deductive sciences are altogether inductive, and that their first principles are generalisations from experience. It remains to be examined whether this science resembles geometry in the further circumstance, that some of its inductions are not exactly true; and that the peculiar certainty ascribed to it, on account of which its propositions are called Necessary Truths, is fictitious and hypothetical, being true in no other sense than that those propositions necessarily follow from the hypothesis of the truth of premisses which are avowedly mere approximations to truth.

Notes

1 Unless, indeed, the consequent was generated not by the antecedent, but by the means we employed to produce the antecedent. As, however, these means are under our power, there is so far a probability that they are also sufficiently within our knowledge, to enable us to judge whether that could be the case or not.
2 *Discourse on the Study of Natural Philosophy*, p. 179.
3 WHEWELL'S *Philosophy of the Inductive Sciences*, i., 59–61.
4 Ibid., 57.
5 *Philosophy of the Inductive Sciences*, i., 54, 55.
6 It would be difficult to name a man more remarkable at once for the greatness and the universality of his intellectual powers, than Leibnitz. Yet this great man gave as a reason for rejecting Newton's scheme of the solar system, that God *could not* make a body revolve round a distant centre, unless either by some impelling mechanism, or by miracle: – "Tout ce qui nest pas explicable" says he in a letter to the Abbé Conti, "par la nature des créatures, est miraculeux. Il ne suffit pas de dire: Dieu a fait une telle loi de nature; donc la chose est naturelle. Il faut que la loi soit exécutable par les natures des créatures. Si Dieu donnait cette loi, par exemple, à un corps libre, de tourner à l'entour d'un certain centre, *il faudrait ou qu'il y joignît d'autres corps qui par leur impulsion l'obligeassent de rester toujours dans son orbite circulaire, ou qu'il mît un ange à ses trousses, ou enfin il faudrait qu'il y concourût extraordinairement;* car naturellement il s'écartera par la tangente." – *Works of Leibnitz*, ed. Dutens, iii. 446.
7 *Philosophy of the Inductive Sciences,* vol. ii., p. 174.
8 *Philosophy of the Inductive Sciences,* i., 238.
9 *Ibid.,* 237.
10 *Philosophy of the Inductive Sciences*, i., 213.
11 *Ibid.*, 384, 385.
12 See, for illustrations of various sorts, Professor LESLIE'S *Philosophy of Arithmetic;* and see also two of the most efficient books ever written for training the infant intellect, Mr. HORACE GRANT'S *Arithmetic for Young Children,* and his *Second Stage of Arithmetic*, both published by the Society for the Diffusion of Useful Knowledge.

"Number," says the reviewer of Mr. Whewell, already cited, "we cannot help regarding as an abstraction, and consequently its general properties or its axioms to be of necessity inductively concluded from the consideration of particular cases. And surely this is the way in which children do acquire their knowledge of number, and in which they learn its axioms. The apples and the marbles are put in requisition, and through the multitude of gingerbread nuts their ideas acquire clearness, precision, and generality."

Part 5

POSITIVISM

Positivism, now an often maligned and misunderstood philosophical doctrine, was also born over the course of the nineteenth century. Though it would be inaccurate to call him a positivist, the basic idea can already be found in the works of Mill and his emphasis on empiricism, as we saw in the previous excerpts. Reacting against the *a priori* turn they saw as exemplified by philosophers like Kant and his followers, positivists turned back toward a perspective that privileges, in one way or another, our direct sensory experience. All knowledge must be either *a priori* and thus linguistic (definitional or conventional), or derived from sense experience.

There are manifold ways in which to expand and elaborate upon this relatively simple aphorism, and we will encounter three of them in this part of the volume. A positivist picture of the world will naturally focus closely on the sciences, as these are the places where human observational powers are most systematically refined and honed. One can also see an affinity for a distrust of theoretical reasoning (being more remote from our immediate experience), as well as for an anti-realist approach to scientific knowledge – that is, the claim that science does not give us access to facts about the underlying ontology of the world, but rather only to descriptions of the observable states of the world and predictions of future such observable states.

Far from being an esoteric metaphysical debate, worries about the proper epistemic attitude that we ought to have with respect to claims in the sciences became significantly more acute over the course of the nineteenth century. While the relationship between empirical observation and scientific theory had already been strained by the postulation of Newtonian forces, as we have seen in this volume, nineteenth-century scientists would go on to add to the realm of apparently legitimate scientific study matters as diverse as fields, fossils from the distant past, chemical elements, and properties of human minds and societies. If our science is to successfully make reference to these kinds of entities, we need an understanding of the nature of scientific knowledge that is sufficiently complex to let us explain how we have epistemic access to them, and what that epistemic access looks like. Viewed in this context, positivism is a reasonable response to the state of scientific advancement, an effort to find an indisputable foundation for scientific knowledge grounded in the kind of direct observation of the external world which could, plausibly enough, be taken to be fundamental, and from which we could build intersubjective, community agreement.

DOI: 10.4324/9781003263814-15

Editorial Headnote

11. Auguste Comte, *The Positive Philosophy of Auguste Comte*, tr. Harriet Martineau (1830, tr. 1853)

As a philosophical movement, positivism finds its origin in the "positive philosophy" of the French thinker Auguste Comte (1798–1857), most especially his *Cours de philosophie positive* (Course of Positive Philosophy) published in French in 1830. Comte's theory centers around a three-phase development of human knowledge which, he argues, can be viewed in the history of every major branch of study: first through a *theological* phase, which ascribes causes to the direct action of supernatural beings or powers; then a *metaphysical* phase, which converts those supernatural beings into forces inherent in the objects we study in nature; and finally a *scientific* or *positive* phase, which recognizes that underlying knowledge of such forces is unattainable, and we should rest content with knowledge of the general laws that describe the behavior of the systems that we wish to study. The theological stage must, Comte argues, be the first, for it offers us the only way to observe facts without already being in possession of a theory – by analogy with our own ability to act in the world.[1]

It is the discovery of quantitative, general laws to which science should dedicate itself in the final, positive phase. Newton's law of gravity provides a paradigmatic example: it explains the behavior of bodies under gravitation perfectly, as well as offering grounds for predictions of future trajectories. But it should not, Comte argues, be interpreted as "explaining" those movements in the sense of revealing the truth of some kind of real, underlying cause or force that moves those bodies (such a force would belong to the metaphysical, not the positive stage of development). Physics, astronomy, chemistry, and even physiology are now positive sciences, he argues; the social sciences are too complex to have yet entered into their positive phase. Among Comte's aims is to develop some first outlines toward such a positive philosophy of social phenomena. His emphasis on a general scientific knowledge that can transcend particular scientific disciplines explains his development of a broad, philosophical perspective on science.

The translator of this work into English (who also condensed and edited the work for a British audience), Harriet Martineau (1802–1876), was a crucial figure in the development of English economics and social theory, as well as a key figure in the abolitionist movement, visiting the United States to speak against slavery in the mid-1830s. She was a close friend of the Darwin family, and went on to publish a number of works supporting a secular philosophy.

Note

1 Larry Laudan has noted that this is an anticipation of the contemporary position of the "theory-ladenness of observation"; the idea that observations necessarily bear the mark of the theory that informs them. See L. Laudan, 'Towards a Reassessment of Comte's "Méthode Positive",' *Philosophy of Science*, 38:1 (1974), pp. 35–53.

11

THE POSITIVE PHILOSOPHY OF AUGUSTE COMTE

Auguste Comte

Source: Auguste Comte, *The Positive Philosophy of Auguste Comte*, tr. Harriet Martineau (1853 [1830]), pp. 25–38.

POSITIVE PHILOSOPHY.

INTRODUCTION.

CHAPTER I.

ACCOUNT OF THE AIM OF THIS WORK. – VIEW OF THE NATURE AND IMPORTANCE OF THE POSITIVE PHILOSOPHY.

A GENERAL statement of any system of philosophy may be either a sketch of a doctrine to be established, or a summary of a doctrine already established. If greater value belongs to the last, the first is still important, as characterizing from its origin the subject to be treated. In a case like the present, where the proposed study is vast and hitherto indeterminate, it is especially important that the field of research should be marked out with all possible accuracy. For this purpose, I will glance at the considerations which have originated this work, and which will be fully elaborated in the course of it.

In order to understand the true value and character of the Positive Philosophy, we must take a brief general view of the progressive course of the human mind, regarded as a whole; for no conception can be understood otherwise than through its history.

Law of human progress.

From the study of the development of human intelligence, in all directions, and through all times, the discovery arises of a great fundamental law, to which it is necessarily subject, and which has a solid foundation of proof, both in the facts of our organization and in our historical experience. The law is this: – that each of our leading conceptions – each branch of our knowledge – passes successively through three different theoretical condititions: the Theological, or fictitious; the Metaphysical, or abstract; and the Scientific, or positive. In other words, the human mind, by its nature, employs in its progress

DOI: 10.4324/9781003263814-16

three methods of philosophizing, the character of which is essentially different, and even radically opposed: viz., the theological method, the metaphysical, and the positive. Hence arise three philosophies, or general systems of conceptions on the aggregate of phenomena, each of which excludes the others. The first is the necessary point of departure of the human understanding; and the third is its fixed and definite state. The second is merely a state of transition.

First Stage.

In the theological state, the human mind, seeking the essential nature of beings, the first and final causes (the origin and purpose) of all effects – in short, Absolute knowledge – supposes all phenomena to be produced by the immediate action of supernatural beings.

Second Stage.

In the metaphysical state, which is only a modification of the first, the mind supposes, instead of supernatural beings, abstract forces, veritable entities (that is, personified abstractions) inherent in all beings, and capable of producing all phenomena. What is called the explanation of phenomena is, in this stage, a mere reference of each to its proper entity.

Third Stage.

In the final, the positive state, the mind has given over the vain search after Absolute notions, the origin and destination of the universe, and the causes of phenomena, and applies itself to the study of their laws – that is, their invariable relations of succession and resemblance. Reasoning and observation, duly combined, are the means of this knowledge. What is now understood when we speak of an explanation of facts is simply the establishment of a connection between single phenomena and some general facts, the number of which continually diminishes with the progress of science.

Ultimate point of each.

The Theological system arrived at the highest perfection of which it is capable when it substituted the providential action of a single Being for the varied operations of the numerous divinities which had been before imagined. In the same way, in the last stage of the Metaphysical system, men substitute one great entity (Nature) as the cause of all phenomena, instead of the multitude of entities at first supposed. In the same way, again, the ultimate perfection of the Positive system would be (if such perfection could be hoped for) to represent all phenomena as particular aspects of a single general fact – such as Gravitation, for instance.

The importance of the working of this general law will be established hereafter. At present, it must suffice to point out some of the grounds of it.

Evidences of the law.

There is no science which, having attained the positive stage, does not bear marks of having passed through the others. Some time since it was (whatever it might be) composed, as we can now perceive, of metaphysical abstractions; and, further back in the course of time, it took its form from theological conceptions. We shall have only too much occasion to see, as we proceed, that our most advanced sciences still bear very evident marks of the two earlier periods through which they have passed.

Actual.

The progress of the individual mind is not only an illustration, but an indirect evidence of that of the general mind. The point of departure of the individual and of the race being the same, the phases of the mind of a man correspond to the epochs of the mind of the race. Now, each of us is aware, if he looks back upon his own history, that he was a theologian in his childhood, a metaphysician in his youth, and a natural philosopher in his manhood. All men who are up to their age can verify this for themselves.

Besides the observation of facts, we have theoretical reasons in support of this law.

Theoretical.

The most important of these reasons arises from the necessity that always exists for some theory to which to refer our facts, combined with the clear impossibility that, at the outset of human knowledge, men could have formed theories out of the observation of facts. All good intellects have repeated, since Bacon's time, that there can be no real knowledge but that which is based on observed facts. This is incontestable, in our present advanced stage; but, if we look back to the primitive stage of human knowledge, we shall see that it must have been otherwise then. If it is true that every theory must be based upon observed facts, it is equally true that facts can not be observed without the guidance of some theory. Without such guidance, our facts would be desultory and fruitless; we could not retain them: for the most part we could not even perceive them.

Thus, between the necessity of observing facts in order to form a theory, and having a theory in order to observe facts, the human mind would have been entangled in a vicious circle, but for the natural opening afforded by Theological conceptions. This is the fundamental reason for the theological character of the primitive philosophy. This necessity is confirmed by the perfect suitability of the theological philosophy to the earliest researches of the human mind. It is remarkable that the most inaccessible questions – those of the nature of beings, and the origin and purpose of phenomena – should be the first to occur in a primitive state, while those which are really within our reach are regarded as almost unworthy of serious study. The reason is evident enough: – that experience alone can teach us the measure of our powers; and if men had not begun by an exaggerated estimate of what they can do, they would never have done all that they are capable of. Our organization requires this. At such a period there could have been no reception of a positive philosophy, whose function is to discover the laws of phenomena, and whose leading characteristic it is to regard as interdicted to human reason those sublime mysteries which theology explains, even to their minutest details, with the most attractive facility. It is just so under a practical view of the nature of the researches with which men first occupied themselves. Such inquiries offered the powerful charm of unlimited empire over the external world – a world destined wholly for our use, and involved in every way with our existence. The theological philosophy, presenting this view, administered exactly the stimulus necessary to incite the human mind to the irksome labor without which it could make no progress. We can now scarcely conceive of such a state of things, our reason having become sufficiently mature to enter upon laborious scientific researches, without needing any such stimulus as wrought upon the imaginations of astrologers and alchemists. We have motive enough in the hope of discovering the laws of phenomena, with a view to the confirmation or rejection of a theory. But it could not be so in the earliest days; and it is to the chimeras of astrology and alchemy that we owe the long series of observations and experiments on which our positive science is based. Kepler felt this on behalf of astronomy, and Berthollet on behalf of chemistry. Thus was a spontaneous philosophy, the theological, the only possible beginning, method, and provisional system, out of which the Positive philosophy could grow. It is easy, after this, to perceive how Metaphysical methods and doctrines must have afforded the means of transition from the one to the other.

The human understanding, slow in its advance, could not step at once from the theological into the positive philosophy. The two are so radically opposed, that an intermediate system of conceptions has been necessary to render the transition possible. It is only in doing this, that metaphysical conceptions have any utility whatever. In contemplating phenomena, men substitute for supernatural direction a corresponding entity. This entity may have been supposed to be derived from the supernatural action: but it is more easily lost sight of, leaving attention free from the facts themselves, till, at length, metaphysical agents have ceased to be anything more than the abstract names of phenomena. It is not easy to say by what other process than this our minds could have passed from supernatural considerations to natural; from the theological system to the positive.

The law of human development being thus established, let us consider what is the proper nature of the Positive Philosophy.

Character of the Positive Philosophy.

As we have seen, the first characteristic of the Positive Philosophy is that it regards all phenomena as subjected to invariable natural *Laws*. Our business is, – seeing how vain is any research into what are called *Causes,* whether first or final, – to pursue an accurate discovery of these Laws, with a view to reducing them to the smallest possible number. By speculating upon causes, we could solve no difficulty about origin and purpose. Our real business is to analyse accurately the circumstances of phenomena, and to connect them by the natural relations of succession and resemblance. The best illustration of this is in the case of the doctrine of Gravitation. We say that the general phenomena of the universe are *explained* by it, because it connects under one head the whole immense variety of astronomical facts; exhibiting the constant tendency of atoms toward each other in direct proportion to their masses, and in inverse proportion to the squares of their distance; while the general fact itself is a mere extension of one which is perfectly familiar to us, and which we therefore say that we know; – the weight of bodies on the surface of the earth. As to what weight and attraction are, we have nothing to do with that, for it is not a matter of knowledge at all. Theologians and metaphysicians may imagine and refine about such questions; but positive philosophy rejects them. When any attempt has been made to explain them, it has ended only in saying that attraction is universal weight, and that weight is terrestrial attraction: that is, that the two orders of phenomena are identical; which is the point from which the question set out. Again, M. Fourier, in his fine series of researches on Heat, has given us all the most important and precise laws of the phenomena of heat, and many large and new truths, without once inquiring into its nature, as his predecessors had done when they disputed about calorific matter and the action of a universal ether. In treating his subject in the Positive method, he finds inexhaustible material for all his activity of research, without betaking himself to insoluble questions.

History of the Positive Philosophy.

Before ascertaining the stage which the Positive Philosophy has reached, we must bear in mind that the different kinds of our knowledge have passed through the three stages of progress at different rates, and have not therefore arrived at the same time. The rate of advance depends on the nature of the knowledge in question, so distinctly that, as we shall see hereafter, this consideration constitutes an accessary to the fundamental law of progress. Any kind of knowledge reaches the positive stage early in proportion to its generality, simplicity, and independence of other departments. Astronomical science, which is above all made up of facts that are general, simple, and independent of other sciences, arrived first; then terrestrial Physics; then Chemistry; and, at length, Physiology.

It is difficult to assign any precise date to this revolution in science. It may be said, like everything else, to have been always going on; and especially since the labors of Aristotle and the school of Alexandria; and then from the introduction of natural science into the West of Europe by the Arabs. But, if we must fix upon some marked period, to serve as a rallying point, it must be that, – about two centuries ago, – when the human mind was astir under the precepts of Bacon, the conceptions of Descartes, and the discoveries of Galileo. Then it was that the spirit of the Positive philosophy rose up in opposition to that of the superstitious and scholastic systems which had hitherto obscured the true character of all science. Since that date, the progress of the Positive philosophy, and the decline of the other two, have been so marked that no rational mind now doubts that the revolution is destined to go on to its completion, – every branch of knowledge being, sooner or later, brought within the operation of Positive philosophy. This is not yet the case. Some are still lying outside: and not till they are brought in will the Positive philosophy possess that character of universality which is necessary to its definitive constitution.

New department of Positive philosophy.

In mentioning just now the four principal categories of phenomena, – astronomical, physical, chemical, and physiological, – there was an omission which will have been noticed. Nothing was said of Social phenomena. Though involved with the physiological, Social phenomena demand a distinct classification, both on account of their importance and of their difficulty. They are the most individual, the most complicated, the most dependent on all others; and therefore they must be the latest, – even if they had no special obstacle to encounter. This branch of science has not hitherto entered into the domain of Positive philosophy. Theological and metaphysical methods, exploded in other departments, are as yet exclusively applied, both in the way of inquiry and discussion, in all treatment of Social subjects, though the best minds are heartily weary of eternal disputes about divine right and the sovereignity of the people. This is the great, while it is evidently the only gap which has to be filled, to constitute, solid and entire, the Positive Philosophy. Now that the human mind has grasped celestial and terrestrial physics, – mechanical and chemical; organic physics, both vegetable and animal, – there remains one science, to fill up the series of sciences of observation, – Social physics. This is what men have now most need of: and this it is the principal aim of the present work to establish.

Social Physics.

It would be absurd to pretend to offer this new science at once in a complete state. Others, less new, are in very unequal conditions of forwardness. But the same character of positivity which is impressed on all the others will be shown to belong to this. This once done, the philosophical system of the moderns will be in fact complete, as there will then be no phenomenon which does not naturally enter into some one of the five great categories. All our fundamental conceptions having become homogeneous, the Positive state will be fully established. It can never again change its character, though it will be for ever in course of development by additions of new knowledge. Having acquired the character of universality which has hitherto been the only advantage resting with the two preceding systems, it will supersede them by its natural superiority, and leave to them only an historical existence.

Secondary aim of this work.

We have stated the special aim of this work. Its secondary and general aim is this: – to review what has been effected in the Sciences, in order to show that they are not radically separate, but all branches from the same trunk. If we had confined ourselves to the first and special object of the work, we should have produced merely a study of Social physics: whereas, in introducing the second and general, we offer a study of Positive philosophy, passing in review all the positive sciences already formed.

To review the philosophy of the Sciences.

The purpose of this work is not to give an account of the Natural Sciences. Besides that it would be endless, and that it would require a scientific preparation such as no one man possesses, it would be apart from our object, which is to go through a course of not Positive Science, but Positive Philosophy. We have only to consider each fundamental science in its relation to the whole positive system, and to the spirit which characterizes it; that is, with regard to its methods and its chief results.

The two aims, though distinct, are inseparable; for, on the one hand, there can be no positive philosophy without a basis of social science, without which it could not be all-comprehensive; and, on the other hand, we could not pursue Social science without having been prepared by the study of phenomena less complicated than those of society, and furnished with a knowledge of laws and anterior facts which have a bearing upon social science. Though the fundamental sciences are not all equally interesting to ordinary minds, there is no one of them that can be neglected in an inquiry like the present; and, in the eye of philosophy, all are of equal value to human welfare. Even those which appear the least interesting have their own value, either on account of the perfection of their methods, or as being the necessary basis of all the others.

Speciality.

Lest it should be supposed that our course will lead us into a wilderness of such special studies as are at present the bane of a true positive philosophy, we will briefly advert to the existing prevalence of such special pursuit. In the primitive state of human knowledge there is no regular division of intellectual labor. Every student cultivates all the sciences. As knowledge accrues, the sciences part off; and students devote themselves each to some one branch. It is owing to this division of employment, and concentration of whole minds upon a single department, that science has made so prodigious an advance in modern times; and the perfection of this division is one of the most important characteristics of the Positive philosophy. But, while admitting all the merits of this change, we can not be blind to the eminent disadvantages which arise from the limitation of minds to a particular study. It is inevitable that each should be possessed with exclusive notions, and be therefore incapable of the general superiority of ancient students, who actually owed that general superiority to the inferiority of their knowledge. We must consider whether the evil can be avoided without losing the good of the modern arrangement; for the evil is becoming urgent. We all acknowledge that the divisions established for the convenience of scientific pursuit are radically artificial; and yet there are very few who can embrace in idea the whole of any one science: each science moreover being itself only a part of a great whole. Almost every one is busy about his own particular section, without much thought about its relation to the general system of positive knowledge. We must not be blind to the evil, nor slow in seeking a remedy. We must not forget that this is the weak side of the positive philosophy, by which it may yet be attacked, with some hope of success, by the adherents of the theological and metaphysical systems. As to the remedy, it certainly does not lie in a return to the ancient confusion of pursuits, which would be mere retrogression, if it were possible, which it is not. It lies in perfecting the division of employments itself, – in carrying it one degree higher, – in constituting one more speciality from the study of scientific generalities. Let us have a new class of students, suitably prepared, whose business it shall be to take the respective sciences as they are, determine the spirit of each, ascertain their relations and mutual connection, and reduce their respective principles to the smallest number of general principles, in conformity with the fundamental rules of the Positive Method. At the same time, let other students be prepared for their special pursuit by an education which recognises the whole scope of positive science, so as to profit by the labors of the students of generalities, and so as to correct reciprocally, under that guidance, the results obtained by each. We see some approach already to this arrangement. Once established, there would be nothing to apprehend from any extent of division of employments. When we once have a class of learned men, at the disposal of all others, whose business it shall be to connect each new discovery with the general system, we may dismiss all fear of the great whole being lost sight of in the pursuit of the details of knowledge. The organization of scientific research will then be complete; and it will henceforth have occasion only to extend its development, and not to change its character. After all, the formation of such a new class as is proposed would be merely an extension of the principle which has created all the classes we have. While science was narrow, there was only one class:

Proposed new class of students.

as it expanded, more were instituted. With a further advance a fresh need arises, and this new class will be the result.

Advantages of the Positive Phlosophy.

The general spirit of a course of Positive Philosophy having been thus set forth, we must now glance at the chief advantages which may be derived, on behalf of human progression, from the study of it. Of these advantages, four may be especially pointed out.

Illustrates the Intellectual function.

I. The study of the Positive Philosophy affords the only rational means of exhibiting the logical laws of the human mind, which have hitherto been sought by unfit methods. To explain what is meant by this, we may refer to a saying of M. de Blainville, in his work on Comparative Anatomy, that every active, and especially every living being, may be regarded under two relations – the Statical and the Dynamical; that is, under conditions or in action. It is clear that all considerations range themselves under the one or the other of these heads. Let us apply this classification to the intellectual functions.

If we regard these functions under their Statical aspect – that is, if we consider the conditions under which they exist – we must determine the organic circumstances of the case, which inquiry involves it with anatomy and physiology. If we look at the Dynamic aspect, we have to study simply the exercise and results of the intellectual powers of the human race, which is neither more nor less than the general object of the Positive Philosophy. In short, looking at all scientific theories as so many great logical facts, it is only by the thorough observation of these facts that we can arrive at the knowledge of logical laws. These being the only means of knowledge of intellectual phenomena, the illusory psychology, which is the last phase of theology, is excluded. It pretends to accomplish the discovery of the laws of the human mind by contemplating it in itself; that is, by separating it from causes and effects. Such an attempt, made in defiance of the physiological study of our intellectual organs, and of the observation of rational methods of procedure, can not succeed at this time of day.

The Positive Philosophy, which has been rising since the time of Bacon, has now secured such a preponderance, that the metaphysicians themselves profess to ground their pretended science on an observation of facts. They talk of external and internal facts, and say that their business is with the latter. This is much like saying that vision is explained by luminous objects painting their images upon the retina. To this the physiologists reply that another eye would be needed to see the image. In the same manner, the mind may observe all phenomena but its own. It may be said that a man's intellect may observe his passions, the seat of the reason being somewhat apart from that of the emotions in the brain; but there can be nothing like scientific observation of the passions, except from without, as the stir of the emotions disturbs the observing faculties more or less. It is yet more out of the question to make an intellectual observation of intellectual processes. The observing and observed organ are here the same, and its action can not be pure and natural. In order to observe, your intellect must pause from activity; yet it is this very activity that you want to observe. If you can not effect the pause, you can not observe: if you do effect it, there is nothing to observe. The results of such a method are in proportion to its absurdity. After two thousand years of psychological pursuit, no one proposition is established to the satisfaction of its followers. They are divided, to this day, into a multitude of schools, still disputing about the very elements of their doctrine. This interior observation gives birth to almost as many theories as there are observers. We ask in vain for any one discovery, great or small, which has been made under this method. The psychologists have done some good in keeping up the activity of our understandings, when there was no better work for our faculties to do; and they may have added something to our stock of knowledge. If they have done so, it is by practising the Positive method – by observing the progress of the human mind in the light of science; that is, by ceasing, for the moment, to be psychologists.

The view just given in relation to logical Science becomes yet more striking when we consider the logical Art.

The Positive Method can be judged of only in action. It can not be looked at by itself, apart from the work on which it is employed. At all events, such a contemplation would be only a dead study, which could produce nothing in the mind which loses time upon it. We may talk for ever about the method, and state it in terms very wisely, without knowing half so much about it as the man who has once put it in practice upon a single particular of actual research, even without any philosophical intention. Thus it is that psychologists, by dint of reading the precepts of Bacon and the discourses of Descartes, have mistaken their own dreams for science.

Without saying whether it will ever be possible to establish, *à priori,* a true method of investigation, independent of a philosophical study of the sciences, it is clear that the thing has never been done yet, and that we are not capable of doing it now. We can not, as yet, explain the great logical procedures, apart from their applications. If we ever do, it will remain as necessary then as now to form good intellectual habits by studying the regular application of the scientific methods which we shall have attained.

This, then, is the first great result of the Positive Philosophy – the manifestation by experiment of the laws which rule the Intellect in the investigation of truth; and, as a consequence, the knowledge of the general rules suitable for that object.

Must regenerate Education.

II. The second effect of the Positive Philosophy, an effect not less important and far more urgently wanted, will be to regenerate Education.

The best minds are agreed that our European education, still essentially theological, metaphysical, and literary, must be superseded by a Positive training, conformable to our time and needs. Even the governments of our day have shared, where they have not originated, the attempts to establish positive instruction; and this is a striking indication of the prevalent sense of what is wanted. While encouraging such endeavors to the utmost, we must not however, conceal from ourselves that everything yet done is inadequate to the object. The present exclusive specialty of our pursuits, and the consequent isolation of the sciences, spoil our teaching. If any student desires to form an idea of natural philosophy as a whole, he is compelled to go through each department as it is now taught, as if he were to be only an astronomer, or only a chemist; so that, be his intellect what it may, his training must remain very imperfect. And yet his object requires that he should obtain general positive conceptions of all the classes of natural phenomena. It is such an aggregate of conceptions, whether on a great or on a small scale, which must henceforth be the permanent basis of all human combinations. It will constitute the mind of future generations. In order to this regeneration of our intellectual system, it is necessary that the sciences, considered as branches from one trunk, should yield us, as a whole, their chief methods and their most important results. The specialities of science can be pursued by those whose vocation lies in that direction. They are indispensable; and they are not likely to be neglected; but they can never of themselves renovate our system of Education; and, to be of their full use, they must rest upon the basis of that general instruction which is a direct result of the Positive Philosophy.

Advances sciences by combining them.

III. The same special study of scientific generalities must also aid the progress of the respective positive sciences: and this constitutes our third head of advantages.

The divisions which we establish between the sciences are, though not arbitrary, essentially artificial. The subject of our researches is one: we divide it for our convenience, in order to deal the more easily with its difficulties. But it sometimes happens – and especially with the most important doctrines of each science – that we need what we can not obtain under the present isolation of the sciences – a combination of several special points of view; and for want of this, very important problems wait for their solution much longer than they otherwise need do. To go back into the past for example: Descartes' grand conception with regard to analytical geometry is a discovery which has changed the whole aspect of mathematical science, and yielded the germ of all future progress; and it issued from the union of two sciences which had always before been separately regarded and pursued. The case of pending questions is yet more impressive; as, for instance, in Chemistry, the doctrine of Definite Proportions. Without entering upon the discussion of the fundamental principle of this theory, we may say with assurance that, in order to determine it – in order to determine whether it is a law of nature that atoms should necessarily combine in fixed numbers – it will be indispensable that the chemical point of view should be united with the physiological. The failure of the theory with regard to organic bodies indicates that the cause of this immense exception must be investigated; and such an inquiry belongs as much to physiology as to chemistry. Again, it is as yet undecided whether azote is a simple or a compound body. It was concluded by almost all chemists that azote is a simple body; the illustrious Berzilius hesitated, on purely chemical considerations; but he was also influenced by the physiological observation that animals which receive no azote in their food have as much of it in their tissues as carnivorous animals. From this we see how physiology must unite with chemistry to inform us whether azote is simple or compound, and to institute a new series of researches upon the relation between the composition of living bodies and their mode of alimentation.

Such is the advantage which, in the third place, we shall owe to Positive philosophy – the elucidation of the respective sciences by their combination. In the fourth place

IV. The Positive Philosophy offers the only solid basis for that Social Reorganization which must succeed the critical condition in which the most civilized nations are now living.

Must reorganize society.

It can not be necessary to prove to anybody who reads this work that Ideas govern the world, or throw it into chaos; in other words, that all social mechanism rests upon Opinions. The great political and moral crisis that societies are now undergoing is shown by a rigid analysis to arise out of intellectual anarchy. While stability in fundamental maxims is the first condition of genuine social order, we are suffering under an utter disagreement which may be called universal. Till a certain number of general ideas can be acknowledged as a rallying-point of social doctrine, the nations will remain in a revolutionary state, whatever palliatives may be devised; and their institutions can be only provisional. But whenever the necessary agreement on first principles can be obtained, appropriate institutions will issue from them, without shock or resistance; for the causes of disorder will have been arrested by the mere fact of the agreement. It is in this direction that those must look who desire a natural and regular, a normal state of society.

Now, the existing disorder is abundantly accounted for by the existence, all at once, of three incompatible philosophies – the theological, the metaphysical, and the positive. Any one of these might alone secure some sort of social order; but while the three co-exist, it is impossible for us to understand one another upon any essential point whatever. If this is true, we have only to ascertain which of the philosophies must, in the nature of things, prevail; and, this ascertained, every man, whatever may have been his former views, can not but concur in its triumph. The problem once recognised, can not remain long unsolved; for all considerations whatever point to the Positive Philosophy as the one destined to prevail. It alone has been advancing during a course of centuries, throughout which the others have been declining. The fact is incontestable. Some may deplore it, but none can destroy it, nor therefore neglect it but under penalty of being betrayed by illusory speculations. This general revolution of the human mind is nearly accomplished. We have only to complete the Positive Philosophy by bringing Social phenomena within its comprehension, and afterward consolidating the whole into one body of homogeneous doctrine. The marked preference which almost all minds, from the highest to the commonest, accord to positive knowledge over vague and mystical conceptions, is a pledge of what the reception of this philosophy will be when it has acquired the only quality that it now wants – a character of due generality. When it has become complete, its supremacy will take place spontaneously, and will re-establish order throughout society. There is, at present, no conflict but between the theological and the metaphysical philosophies. They are contending for the task of reorganizing society; but it is a work too mighty for either of them.

The positive philosophy has hitherto intervened only to examine both, and both are abundantly discredited by the process. It is time now to be doing something more effective, without wasting our forces in needless controversy. It is time to complete the vast intellectual operation begun by Bacon, Descartes, and Galileo, by constructing the system of general ideas which must henceforth prevail among the human race. This is the way to put an end to the revolutionary crisis which is tormenting the civilized nations of the world.

Leaving these four points of advantage, we must attend to one precautionary reflection.

No hope of reduction to a single law.

Because it is proposed to consolidate the whole of our acquired knowledge into one body of homogeneous doctrine, it must not be supposed that we are going to study this vast variety as proceeding from a single principle, and as subjected to a single law. There is something so chimerical in attempts at universal explanation by a single law, that it may be as well to secure this Work at once from any imputation of the kind, though its development will show how undeserved such an imputation would be. Our intellectual resources are too narrow, and the universe is too complex, to leave any hope that it will ever be within our power to carry scientific perfection to its last degree of simplicity. Moreover, it appears as if the value of such an attainment, supposing it possible, were greatly overrated. The only way, for instance, in which we could achieve the business, would be by connecting all natural phenomena with the most general law we know – which is that of gravitation, by which astronomical phenomena are already connected with a portion of terrestrial physics. Laplace has indicated that chemical phenomena may be regarded as simple atomic effects of the Newtonian attraction, modified by the form and mutual position of the atoms. But supposing this view proveable (which it can not be while we are without data about the constitution of bodies), the difficulty of its application would doubtless be found so great that we must still maintain the existing division between astronomy and chemistry, with the difference that we now regard as natural that division which we should then call artificial. Laplace himself presented his idea only as a philosophic device, incapable of exercising any useful influence over the progress of chemical science. Moreover, supposing this insuperable difficulty overcome, we should be no nearer to scientific unity, since we then should still have to connect the whole of physiological phenomena with the same law, which certainly would not be the least difficult part of the enterprise. Yet, all things considered, the hypothesis we have glanced at would be the most favorable to the desired unity.

The consideration of all phenomena as referable to a single origin is by no means necessary to the systematic formation of science, any more than to the realization of the great and happy consequences that we anticipate from the positive philosophy. The only necessary unity is that of Method, which is already in great part established. As for the doctrine, it need not be *one;* it is enough that it be *homogeneous.* It is, then, under the double aspect of unity of method and homogeneousness of doctrine that we shall consider the different classes of positive theories in this work.

While pursuing the philosophical aim of all science, the lessening of the number of general laws requisite for the explanation of natural phenomena, we shall regard as presumptuous every attempt, in all future time, to reduce them rigorously to one.

Having thus endeavored to determine the spirit and influence of the Positive Philosophy, and to mark the goal of our labors, we have now to proceed to the exposition of the system; that is, to the determination of the universal, or encyclopædic order, which must regulate the different classes of natural phenomena, and consequently the corresponding positive sciences.

Editorial Headnote

12. Ernst Mach, "Introductory Remarks: Antimetaphysical," from *The Analysis of Sensations*, tr. C. M. Williams (1897)

To see the continued trajectory of positivism as an approach to the sciences, we need to move late into the nineteenth century, to Ernst Mach (1838–1916), one of the founding fathers of German positivism. His work was widely circulated in Britain upon publication, and he maintained an extensive correspondence network, including with Karl Pearson (author of the next selection). Mach's work in physics and physiology – the latter being the subject of the work excerpted here – was grounded in a positivist view of the world.

We saw in Comte's "positive stage" of scientific development that, for positivists, the aim of science is not to describe the causes or forces underlying change, but rather to understand the laws that both describe and enable us to predict that change. Mach goes one step further, grounding this role for description in the nature of our mental faculties. We have two habits of mind, "apt to come into strange conflict" (pp. 212–3) with one another: on the one hand, to recognize certain objects as permanent over time, even as they change (e.g., my desk, even as it becomes stained or damaged), and on the other, to break down objects into their component qualities (e.g., the flatness, color, weight, and so forth of the desk). For Mach, objects are *nothing more* than these collections of qualities. Science, then, is not aimed at telling us what these objects "really are" – this is a nonsensical question, as we have access only to the different collections of perceptions that we have of those objects over time.

This approach has a number of important consequences. First, because our only access to objects is through our perceptions of their qualities, any investigation of scientific notions like space or time will necessarily be not only an investigation of the external world, but also of the ways in which some apparently "internal" or "mental" aspects of our perception interact with one another and with those aspects of the "external" world. The concept of "an antithesis of body and spirit" (p. 215) is to be set aside, despite its intuitive appeal; human perceptual and physiological study becomes an essential aspect of scientific methodology, rather than some separate or dualist domain of "the mind." (We will see similar discussions in the third volume, in the methodology of psychology.)

Second, because even in the physical sciences "bodies are but thought-symbols for complexes of sensations (complexes of elements)" (p. 219), the goal of science is thus to create novel "thought-elements, which are able to represent [a given] field [of experience]" (p. 220). Good scientific inquiry is not good at all times and for all purposes. Rather, good science is that which "accommodates itself, with the least expenditure of energy, that is, more economically than any other, to the present *temporary collective state of knowledge*" (p. 220, orig. emph.). Put differently, we do science in order to economize, simplify, and make useful our mental representations for collections of perceptions and their changes over time. Mathematical formulas, for instance, are able in a quick shorthand to describe a massive variety of possible changes in a system; this is what makes Newton's laws so exceptionally important.

Finally, this perspective – and here Mach departs from Comte – does not imply that "metaphysical" views, such as those that would emphasize a unity of the ego or the self, or those that would argue for the existence of causal connections beyond our perceptions, are useless or somehow inherently contradictory. Those views were suitable for certain ends and certain states of collective knowledge, and we return to them as soon as we stop engaging in philosophical reflection. That must mean, for Mach, that any replacement, positivist view has a comparatively "insignificant and ephemeral" track record (p. 221). The intent behind developing a positivist foundation for science, rather, is to note "*why* and to what *purpose* for the greatest portion of life we hold [our naive views], and *why* and for what *purpose* we are provisorily advised to abandon [them]" (p. 221).

12

INTRODUCTORY REMARKS: ANTIMETAPHYSICAL FROM *THE ANALYSIS OF SENSATIONS*

Ernst Mach

Source: Ernst Mach, "Introductory Remarks: Antimetaphysical," from *The Analysis of Sensations*, 1st edn, tr. C. M. Williams (1897), pp. 1–26.

1.

THE splendid success achieved by physical science in modern times, a success which is not restricted to its own sphere but embraces that of other sciences which employ its help, has brought it about that physical ways of thinking and physical modes of procedure enjoy on all hands unwonted prominence, and that the greatest expectations are associated with their employment. In keeping with this drift of modern inquiry, the physiology of the senses, gradually leaving the paths which were opened by men like Goethe, Schopenhauer, and others, but with particular success by Johannes Müller, has also assumed an almost exclusively physical character. This tendency must appear to us as not exactly the proper one, when we reflect that physics despite its considerable development nevertheless constitutes but a portion of a *larger* collective body of knowledge, and that it is unable, with its limited intellectual implements, created for limited and special purposes, to exhaust all the subject-matter of science. Without renouncing the support of physics, it is possible for the physiology of the senses, not only to pursue its own course of development, but also to afford to physical science itself powerful assistance; a point which the following simple considerations will serve to illustrate.

2.

Colors, sounds, temperatures, pressures, spaces, times, and so forth, are connected with one another in manifold ways; and with them are associated moods of mind, feelings, and volitions. Out of this fabric, that which is relatively more fixed and permanent stands prominently forth, engraves itself in the memory, and expresses itself in language. Relatively greater permanency exhibit, first, certain *complexes* of colors, sounds, pressures, and so forth, connected in time and space,

DOI: 10.4324/9781003263814-17

which therefore receive special names, and are designated *bodies*. Absolutely permanent such complexes are not.

My table is now brightly, now dimly lighted. Its temperature varies. It may receive an ink stain. One of its legs may be broken. It may be repaired, polished, and replaced part for part. But for me, amid all its changes, it remains the table at which I daily write.

My friend may put on a different coat. His countenance may assume a serious or a cheerful expression. His complexion, under the effects of light or emotion, may change. His shape may be altered by motion, or be definitely changed. Yet the number of the permanent features presented, compared with the number of the gradual alterations, is always so great, that the latter may be overlooked. It is the same friend with whom I take my daily walk.

My coat may receive a stain, a tear. My very manner of expression shows that we are concerned here with a sum-total of permanency, to which the new element is added and from which that which is lacking is subsequently taken away.

Our greater intimacy with this sum-total of permanency, and its preponderance as contrasted with the changeable, impel us to the partly instinctive, partly voluntary and conscious economy of mental representation and designation, as expressed in ordinary thought and speech. That which is perceptually represented in a single image receives *a single* designation, *a single* name.

As relatively permanent, is exhibited, further, that complex of memories, moods, and feelings, joined to a particular body (the human body), which is denominated the "I" or "Ego." I may be engaged upon this or that subject, I may be quiet or animated, excited or ill-humored. Yet, pathological cases apart, enough durable features remain to identify the ego. Of course, the ego also is only of relative permanency.[1]

After a first survey has been obtained, by the formation of the substance-concepts "body" and "ego" (matter and soul), the will is impelled to a more exact examination of the *changes* that take place in these relatively permanent existences. The changeable features of bodies and of the ego, in fact, are exactly what moves the will to this examination. Here the component parts of the complex are first exhibited as its properties. A fruit is sweet; but it can also be bitter. Also, other fruits may be sweet. The red color we are seeking is found in many bodies. The neighborhood of some bodies is pleasant; that of others, unpleasant. Thus, gradually, different complexes are found to be made up of common elements. The visible, the audible, the tangible, are separated from bodies. The visible is analysed into colors and into form. In the manifoldness of the colors, again, though here fewer in number, other component parts are discerned – such as the primary colors, and so forth. The complexes are disintegrated into *elements*.

3.

The useful habit of designating such relatively permanent compounds by *single* names, and of apprehending them by *single* thoughts, without going to the trouble each time of an analysis of their component parts, is apt to come into strange conflict with the tendency to isolate the component parts. The vague image which we have of a given permanent complex, being an image which does not perceptibly change when one or another of the component parts is taken away, gradually establishes itself as something which exists *by itself.* Inasmuch as it is possible to take away *singly* every constituent part without destroying the capacity of the image to *stand for* the totality and of being recognised again, it is imagined that it is possible to subtract *all* the parts and to have something still remaining. Thus arises the monstrous notion of a *thing in itself,* unknowable and different from its "phenomenal" existence.

Thing, body, matter, are nothing apart from their complexes of colors, sounds, and so forth – nothing apart from their so-called attributes. That Protean, supposititious problem, which springs up so much in philosophy, of a *single* thing with *many* attributes, arises wholly from a mistaking of the fact, that summary comprehension and precise analysis, although both are provisionally justifiable and for many purposes profitable, cannot and must not be carried on *simultaneously.* A body is one and unchangeable only so long as it is unnecessary to consider its details. Thus both the earth and a billiard-ball are spheres, if the purpose in hand permits our neglecting deviations from the spherical form, and great precision is not necessary. But when we are obliged to carry on investigations in orography or microscopy, both bodies cease to be spheres.

4.

Man possesses, in its highest form, the power of consciously and arbitrarily determining his point of view. He can at one time disregard the most salient features of an object, and immediately thereafter give attention to its smallest details; now consider a stationary current, without a thought of its contents, and then measure the width of a Fraunhofer line in the spectrum; he can rise at will to the most general abstractions or bury himself in the minutest particulars. The animal possesses this capacity in a far less degree. It does not assume a point of view, but is usually forced to it. The babe who does not know its father with his hat on, the dog that is perplexed at the new coat of its master, have both succumbed in this conflict of points of view. Who has not been worsted in similar plights? Even the man of philosophy at times succumbs, as the grotesque problem, above referred to, shows.

In this last case, the circumstances appear to furnish a real ground of justification. Colors, sounds, and the odors of bodies are evanescent. But the tangible part, as a sort of constant, durable nucleus, not readily susceptible of annihilation, remains behind; appearing as the vehicle of the more fugitive properties annexed to it. Habit, thus, keeps our thought firmly attached to this central nucleus, even where the knowledge exists that seeing, hearing, smelling, and *touching* are intimately akin in character. A further consideration is, that owing to the singularly extensive development of mechanical physics a kind of *higher reality* is ascribed to space and time than to colors, sounds, and odors; agreeably to which, the temporal and spatial *links* of colors, sounds, and odors appear to be *more real* than the colors, sounds, and odors themselves. The physiology of the senses, however, demonstrates, that spaces and times may just as appropriately be called sensations as colors and sounds.

5.

The ego, and the relation of bodies to the ego, give rise to similar pseudo-problems, the character of which may be briefly indicated as follows:

Let those complexes of colors, sounds, and so forth, commonly called bodies, be designated, for the sake of simplicity, by *A B C* . . .; the complex, known as our own body, which constitutes a part of the former, may be called *K L M* . . .; the complex composed of volitions, memory-images, and the rest, we shall represent by *α β γ* . . . Usually, now, the complex *α β γ* . . . *K L M* . . ., as making up the ego, is opposed to the complex *A B C* . . ., as making up the world of substance; sometimes, also, *α β γ* . . . is viewed as ego, and *K L M* . . *A B C* . . . as world of substance. Now, at first blush, *A B C* . . . appears independent of the ego, and opposed to it as a separate existence. But this independence is only relative, and gives way upon closer inspection. Much, it is true, may change in the complex *α β γ* . . . without a perceptible change being induced in *A B C* . . .; and *vice versa.* But many changes in *α β γ* . . . do pass, by way of changes in *K L M* . . ., to *A B C* . . .; and *vice versa.* (As, for example, when powerful ideas burst forth into acts, or our environment induces noticeable changes in our body.) At the same time the group *K L M* . . . appears to be more intimately connected with *α β γ* . . . and with *A B C* . . ., than the latter do with one another; relations which find their expression in common thought and speech.

Precisely viewed, however, it appears that the group *A B C* . . . is *always* codetermined by *K L M.* A cube of wood when seen close at hand, looks large; when seen at a distance, small; it looks different with the right eye from what it does with the left; sometimes it appears double; with closed eyes it is invisible. The properties of the same body, therefore, appear modified by our own body; they appear conditioned by it. But where, now, is that *same* body, which to the appearance is so *different?* All that can be said is, that with different *K L M* different *A B C* . . . are associated.[2]

We see an object having a point *S.* If we touch *S,* that is, bring it into connexion with our body, we receive a prick. We can see *S,* without feeling the prick. But as soon as we feel the prick we find *S.* The visible point, therefore, is a *permanent fact* or *nucleus,* to which the prick is annexed, according to circumstances, as something accidental. From the frequency of such occurrences we ultimately accustom ourselves to regard *all* properties of bodies as "effects" proceeding from permanent nuclei and conveyed to the ego through the medium of the body; which effects we call *sensations.* By this operation, however, our imagined nuclei are deprived of their entire sensory contents, and converted into mere mental symbols. The assertion, then, is correct that the world consists only of our sensations. In which case we have knowledge *only* of sensations, and the assumption of the nuclei referred to, or of a reciprocal action between them, from which sensations proceed, turns out to be quite idle and superfluous. Such a view can only suit with a half-hearted realism or a half-hearted philosophical criticism.

6.

Ordinarily the complex $\alpha\,\beta\,\gamma\ldots K\,L\,M\ldots$ is contrasted as ego with the complex $A\ B\ C$. Those elements only of $A\ B\ C\ldots$ that more strongly alter $\alpha\,\beta\,\gamma\ldots$, as a prick, a pain, are wont to be comprised in the ego. Afterwards, however, through observations of the kind just referred to, it appears that the right to annex $A\ B\ C\ldots$ to the ego nowhere ceases. In conformity with this view the ego can be so extended as ultimately to embrace the entire world.[3] The ego is not sharply marked off, its limits are very indefinite and arbitrarily displaceable. Only by failing to observe this fact, and by unconsciously narrowing those limits, while at the same time we enlarge them, arise, in the conflict of points of view, the metaphysical difficulties met with in this connexion.

As soon as we have perceived that the supposed unities "body" and "ego" are only makeshifts, designed for provisional survey and for certain practical ends (so that we may take hold of bodies, protect *ourselves* against pain, and so forth), we find ourselves obliged, in many profound scientific investigations, to abandon them as insufficient and inappropriate. The antithesis of ego and world, sensation (phenomenon) and thing, then vanishes, and we have simply to deal with the *connexion* of the *elements* $\alpha\,\beta\,\gamma\ldots A\ B\ C\ldots K\,L\,M\ldots$, of which this antithesis was only a partially appropriate and imperfect expression. This connexion is nothing more nor less than the combination of the above-mentioned elements with other similar elements (time and space). Science has simply to *accept* this connexion, and to set itself aright (get its bearings) in the intellectual environment which is thereby furnished, without attempting to explain its existence.

On a superficial examination the complex $\alpha\,\beta\,\gamma\ldots$ appears to be made up of much more evanescent elements than $A\ B\ C\ldots$ and $K\,L\,M\ldots$ in which last the elements seem to be connected with greater *stability* and *in a more permanent manner* (being joined to solid nuclei as it were). Although on closer inspection the elements of all complexes prove to be *homogeneous,* yet in spite of the knowledge of this fact, the early notion of an antithesis of body and spirit easily regains the ascendancy in the mind. The philosophical spiritualist is often sensible of the difficulty of imparting the needed solidity to his mind-created world of bodies; the materialist is at a loss when required to endow the world of matter with sensation. The *monistic* point of view, which artificial reflexion has evolved, is easily clouded by our older and more powerful instinctive notions.

7.

The difficulty referred to is particularly felt in the following case. In the complex *A B C* . . ., which we have called the world of matter, we find as parts, not only our own body *K L M* . . ., but also the bodies of other persons (or animals) *K'L'M'* . . ., *K" L" M"* . . ., to which, by analogy, we imagine other $\alpha'\beta'\gamma'$ . . ., $\alpha''\beta''\gamma''$ . . ., annexed, similar to $\alpha\,\beta\,\gamma$ . . . So long as we deal with *K'L'M'* . . ., we find ourselves in a thoroughly familiar province at every point sensorially accessible to us. When, however, we inquire after the sensations or feelings appurtenant to the body *K'L'M'* . . ., we no longer find the elements we seek in the province of sense: *we add them in thought.* Not only is the domain which we now enter far less familiar to us, but the transition into it is also relatively unsafe. We have the feeling as if we were plunging into an abyss.[4] Persons who adopt this method only, will never thoroughly rid themselves of this sense of insecurity, which is a frequent source of illusive problems.

But we are not restricted to this course. Let us consider, first, the reciprocal relations of the elements of the complex *A B C* . . ., without regarding *K L M* . . . (our body). All physical investigations are of this sort. A white bullet falls upon a bell; a sound is heard. The bullet turns yellow before a sodium lamp, red before a lithium lamp. Here the elements (*A B C* . . .) appear to be connected only *with one another* and to be independent of our body (*K L M* . . .). But if we take santonine, the bullet again turns yellow. If we press one eye to the side, we see two bullets. If we close our eyes entirely, we see none at all. If we sever the auditory nerve, no sound is heard. The elements *A B C* . . ., therefore, are not only connected among one another, but also with *K L M.* To this extent, and to this extent *only,* do we call *A B C* . . . *sensations,* and regard *A B C* as belonging to the ego. In this way, accordingly, we do not find the gap between bodies and sensations above described, between what is without and what is within, between the material world and the spiritual world.[5] All elements *A B C* . . ., *K L M* . . . constitute a *single* coherent mass only, in which, when any one element is disturbed, *all* is put in motion; except that a disturbance in *K L M* . . . has a more extensive and profound action than in *A B C.* A magnet in our neighborhood disturbs the particles of iron near it; a falling boulder shakes the earth; but the severing of a nerve sets in motion the *whole* system of elements.[6]

8.

That traditional gulf between physical and psychological research, accordingly, exists only for the habitual stereotyped method of observation. A color is a physical object so long as we consider its dependence upon its luminous source, upon other colors, upon heat, upon space, and so forth. Regarding, however, its dependence upon the retina (the elements *K L M* . . .), it becomes a psychological object, a sensation. Not the subject, but the direction of our investigation, is different in the two domains.

Both in reasoning from the observation of the bodies of other men or animals, to the sensations which they possess, as well as in investigating the influence of our own body upon our own sensations, we must complete observed facts by analogy. This is accomplished with much greater readiness and certainty, when it relates, say, only to nervous processes, which cannot be fully observed in our own bodies – that is, when it is carried out in the more familiar physical domain – than when it is made in connexion with psychical processes. Otherwise there is no essential difference.

9.

The considerations advanced will gain in strength and vividness by a concrete example. Thus, I lie upon my sofa. If I close my right eye, the picture represented in the accompanying cut [Fig. 12.1] is presented to my left eye. In a frame formed by the ridge of my eyebrow, by my nose, and by my moustache, appears a part of my body, so far as visible, with its environment.[7] *My* body differs from other human bodies – beyond the fact that every intense motor idea is immediately expressed by a movement of it, and that its being touched determines more striking changes than contact with other bodies – by the circumstance, that it is only partly seen, and, especially, is seen without a head. If I observe an element *A* within my field of vision, and investigate its connexion with another element *B* within the same field, I step out of the domain of physics into that of physiology or psychology, provided *B,* to use the apposite expression of a friend[8] of mine made upon seeing this drawing, passes through my skin. Reflexions like that for the field of vision may be made with regard to the province of touch and the perceptual domains of the other senses.

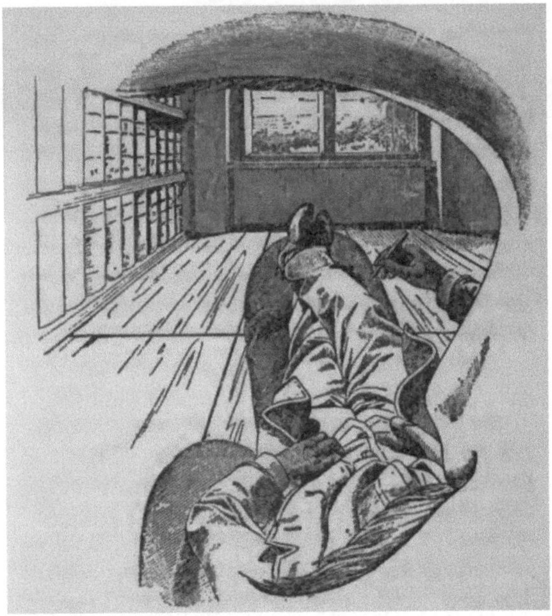

Figure 12.1

10.

Reference has already been made to the different character of the groups of elements designated by $A\ B\ C\ldots$ and $\alpha\ \beta\ \gamma$. As a matter of fact, when we *see* a green tree before us, or *remember* a green tree, that is, represent a green tree to ourselves, we are perfectly aware of the difference of the two cases. The represented tree has a much less determinate, a much more changeable form; its green is much paler and more evanescent; and, what is of especial note, it is plainly situate in a *different* domain. A movement that we *propose* to execute is never more than a represented movement, and appears in a different sphere from that of the executed movement, which always takes place when the image is vivid enough. The statement that the elements A and α appear in different spheres, means, if we go to the bottom of it, simply this, that these elements are united with different other elements. Thus far, therefore, the fundamental constituents of $A\ B\ C\ldots$, $\alpha\ \beta\ \gamma\ldots$ would seem to be *the same* (colors, sounds, spaces, times, motor sensations . . .), and only the character of their connexion different.

Ordinarily pleasure and pain are regarded as different from sensations. Yet not only tactile sensations, but all other kinds of sensations, may pass gradually into pleasure and pain. Pleasure and pain also may be justly termed sensations. Only they are not so well analysed and so familiar as the common sensations. In fact, sensations of pleasure and pain, however faint they may be, really make up the contents of all so-called emotions. Thus, perceptions, ideas, volition, and emotion, in short the whole inner and outer world, are composed of a small number of homogeneous elements connected in relations of varying evanescence or permanence. Usually, these elements are called sensations. But as vestiges of a one-sided theory inhere in that term, we prefer to speak simply of *elements,* as we have already done. The aim of all research is to ascertain the mode of connexion of these elements.[9]

11.

That in this complex of elements, which fundamentally is *one,* the boundaries of bodies and of the ego do not admit of being established in a manner definite and sufficient for all cases, has already been remarked. The comprehending of the elements that are most intimately connected with pleasure and pain, under one ideal mental-economical unity, the ego, is a work of the highest significance for the intellect in the functions which it performs for the pain-avoiding, pleasure-seeking will. The delimitation of the ego, therefore, is instinctively effected, is rendered familiar, and possibly becomes fixed through heredity. Owing to their high practical value, not only for the individual, but for the entire species, the composites "ego" and "body" assert instinctively their claims, and operate with all the power of natural elements. In special cases, however, in which practical ends are not concerned, but where knowledge is an object in itself, the delimitation in question may prove to be insufficient, obstructive, and untenable.[10]

The primary fact is not the *I*, the ego, but the elements (sensations). The elements *constitute* the *I*. *I* have the sensation green, signifies that the element green occurs in a given complex of other elements (sensations, memories). When *I* cease to have the sensation green, when *I* die, then the elements no longer occur in their ordinary, familiar way of association. That is all. Only an ideal mental-economical unity, not a real unity, has ceased to exist.[11]

If a knowledge of the connexion of the elements (sensations) does not suffice us, and we ask, *Who* possesses this connexion of sensations, *Who* experiences the sensations? then we have succumbed to the habit of subsuming every element (every sensation) under some *unanalysed* complex, and we are falling back imperceptibly upon an older, lower, and more limited point of view.[12]

The so-called unity of consciousness is not an argument in point. Since the apparent antithesis of *real* world and *perceived* world is due entirely to our mode of view, and no actual gulf exists between them, a rich and variously interconnected content of consciousness is in no respect more difficult to understand than a rich and diversified interconnexion of the world.

If we regard the ego as a *real* unity, we become involved in the following dilemma: either we must set over against the ego a world of unknowable entities (which would be quite idle and purposeless), or we must regard the whole world, the egos of other people included, as comprised in our own ego (a proposition to which it is difficult to yield serious assent).

But if we take the ego simply as a *practical* unity, put together for purposes of provisional survey, or simply as a more strongly coherent group of elements, less strongly connected with other groups of this kind, questions like those above discussed will not arise and research will have an unobstructed future.

In his philosophical notes Lichtenberg says: "We become conscious of certain percepts that are not dependent upon us; of others that we at least think are dependent upon us. Where is the border-line? We know only the existence of our sensations, percepts, and thoughts. We should say, *It thinks,* just as we say, *It lightens.* It is going too far to say *cogito,* if we translate *cogito* by *I think.* The assumption, or postulation, of the ego is a mere practical necessity." Though the method by which Lichtenberg arrived at this result is somewhat different from ours, we must nevertheless give our full assent to his conclusion.

12.

Bodies do not produce sensations, but complexes of sensations (complexes of elements) make up bodies. If, to the physicist, bodies appear the real, abiding existences, whilst sensations are regarded merely as their evanescent, transitory show, the physicist forgets, in the assumption of such a view, that all bodies are but thought-symbols for complexes of sensations (complexes of elements). Here, too, the *elements* form the real, immediate, and ultimate foundation, which it is the task of physiological research to investigate. By the recognition of this fact, many points of psychology and physics assume more distinct and more economical forms, and many spurious problems are disposed of.

For us, therefore, the world does not consist of mysterious entities, which by their interaction with another, equally mysterious entity, the ego, produce sensations, which alone are accessible. For us, colors, sounds, spaces, times, . . . are the ultimate elements, whose given connexion it is our business to investigate.[13] In this investigation we must not allow our-selves to be impeded by such intellectual abridgments and delimitations as body, ego, matter, mind, etc., which have been formed for special, practical purposes and with wholly provisional and limited ends in view. On the contrary, the fittest forms of thought must be created in and by that research *itself,* just as is done in every special science. In place of the traditional, instinctive ways of thought, a freer, fresher view, conforming to developed experience, must be substituted.

13.

Science always takes its origin in the adaptation of thought to some definite field of experience. The results of the adaptation are thought-elements, which are able to represent the field. The outcome, of course, is different, according to the character and extent of the province surveyed. If the province of experience in question is enlarged, or if several provinces heretofore disconnected are united, the traditional, familiar thought-elements no longer suffice for the extended province. In the struggle of acquired habit with the effort after adaptation, *problems* arise, which disappear when the adaptation is perfected, to make room for others which have arisen in the interim.

To the physicist, *quâ* physicist, the idea of "body" is productive of a real facilitation of view, and is not the cause of disturbance. So, also, the person with purely practical aims, is materially assisted by the idea of the *I* or ego. For, unquestionably, every form of thought that has been designedly or undesignedly constructed for a given purpose, possesses for that purpose a *permanent* value. When, however, research in physics and in psychology meets, the ideas held in the one domain prove to be untenable in the other. From the attempt at mutual adaptation arise the various atomic and monadic theories – which, however, never attain their end. If we regard *sensations,* in the sense above defined, as the *elements of the world,* the problems referred to are practically disposed of, and the *first* and most important adaptation effected. This fundamental view (without any pretension to being a philosophy for all eternity) can at present be adhered to in all provinces of experience; it is consequently the one that accommodates itself with the least expenditure of energy, that is, more economically than any other, to the present *temporary collective state of knowledge.* Furthermore, in the consciousness of its purely economical office, this fundamental view is eminently tolerant. It does not obtrude itself into provinces in which the current conceptions are still adequate. It is ever ready, upon subsequent extensions of the domain of experience, to yield the field to a better conception.

The philosophical point of view of the average man – if that term may be applied to the naïve realism of the ordinary individual – has a claim to the highest consideration. It has arisen in the process of immeasurable time without the conscious assistance of man. It is a product of nature, and is preserved and sustained by nature. Everything that philosophy has accomplished – the *biological* value of every advance, nay, of every error, admitted – is, as compared with it, but an insignificant and ephemeral product of art. The fact is, every thinker, every philosopher, the moment he is forced to abandon his narrow intellectual province by practical necessity, immediately returns to the universal point of view held by all men in common.[14]

To discredit this point of view is not then the purpose of the foregoing "introductory remarks." The task which we have set ourselves is simply to show *why* and to what *purpose* for the greatest portion of life we hold it, and *why* and for what *purpose* we are provisorily obliged to abandon it. No point of view has absolute, *permanent* validity. Each has importance only for some given end.[15]

Notes

1 The apparent permanency of the ego consists chiefly in the fact of its *continuity* and in the slowness of its changes. The many thoughts and plans of yesterday that are continued to-day, and of which our environment in waking hours incessantly reminds us (wherefore in dreams the ego can be very indistinct, doubled, or entirely wanting), and the little habits that are unconsciously and involuntarily kept up for long periods of time, constitute the groundwork of the ego. There can hardly be greater differences in the egos of different people, than occur in the course of years in *one* person. When I recall to-day my early youth, I should take the boy that I then was, with the exception of a few individual features, for a different person, did not the chain of memories that make up my personality lie actually before me. Many an article that I myself penned twenty years ago impresses me now as something quite foreign to myself. The very gradual character of the changes of the body also contributes to the stability of the ego, but in a much less degree than people imagine. Such things are much less analysed and noticed than the intellectual and the moral ego. Personally, people know themselves very poorly.

Once, when a young man, I espied in the street the profile of a face that was very displeasing and repulsive to me. I was not a little taken aback when a moment afterwards I found that it was my own, which, in passing by a place where mirrors were sold, I had perceived reflected from two mirrors that were inclined at the proper angle to each other.

Not long ago, after a trying railway journey by night, and much fatigued, I got into an omnibus, just as another gentleman appeared at the other end. "What degenerate pedagogue is that, that has just entered," thought I. It was myself: opposite me hung a large mirror. The physiognomy of my class, accordingly, was better known to me than my own.

The ego is as little absolutely permanent as are bodies. That which we so much dread in death, the annihilation of our permanency, actually occurs in life in abundant measure. That which is most valued by us, remains preserved in countless copies, or, in cases of exceptional excellence, is even preserved of itself. In the best human being, however, there are individual traits, the loss of which neither he himself nor others need regret. Indeed, at times, death, viewed as a liberation from individuality, may even become a pleasant thought. [When I wrote these lines, Ribot's admirable little book, *The Diseases of Personality,* second edition, Paris, 1883, Chicago, 1895, was unknown to me. Ribot ascribes the principal rôle in preserving the continuity of the ego to the general sensibility. Generally, I am in perfect accord with his views. – Mach, 1895.]

2 A long time ago (in the *Vierteljahrsschrift für Psychiatrie,* Leipsic and Neuwied, 1868, art. "Ueber die Abhängigkeit der Netzhautstellen von einander," I enunciated this thought as follows: The expression "sense-illusion" proves that we are not yet fully conscious, or at least have not yet deemed it necessary to incorporate the fact into our ordinary language, *that the senses represent things neither wrongly nor correctly.* All that can be truly said of the sense-organs is, that, *under different circumstances they produce different sensations and perceptions.* As these "circumstances," now, are extremely manifold in character, being partly external (inherent in the objects), partly internal (inherent in the sensory organs), and partly interior (having their seat in the central organs), it would naturally seem, especially when attention is paid only to external circumstances, as if the organs acted differently under the same conditions. And it is customary to call the unusual effects, deceptions or illusions.

3 When I say that the table, the tree, and so forth, are my sensations, the statement, as contrasted with the mode of representation of the ordinary man, involves a real extension of my ego. On the emotional side also such extensions occur, as in the case of the virtuoso, who possesses as perfect a mastery of his instrument as he does of his own body; or in the case of the skilful orator, on whom the eyes of the audience are all converged, and who is controlling the thoughts of all; or in that of the able politician who is deftly guiding his party; and so on. In conditions of depression, on the other hand such as nervous people often endure, the ego contracts and shrinks. A wall seems to separate it from the world.

4 When I first came to Vienna from the country, as a boy of four or five years, and was taken by my father upon the walls of the city's fortifications, I was very much surprised to see people below in the moat, and could not understand how, from my point of view, they could have got there; for the thought of another way of descent never occurred to me. I remarked the same astonishment, once afterwards in life, in the case of a three-year-old boy of my own, while walking on the walls of Prague. I recall this feeling every time I occupy myself with the reflexion of the text, and I frankly confess that this accidental experience of mine helped to confirm my opinion upon this point, which I have now long held. The habit of pursuing the same methods in material and psychical questions tends greatly to confuse our field of survey. A child, on the piercing of the wall of a house in which it has long dwelt, may experience a veritable enlargement of its world-view, and in the same manner a slight scientific hint may often afford great enlightenment.

5 Compare my *Grundlinien der Lehre von den Bewegungsempfindungen* Leipsic: Engelmann, 1875, p. 54.

6 Quite involuntarily does this relation of things suggest the picture of a viscous mass, at certain places (as in the ego) more firmly coherent than in others. I have often made use of this simile in lectures.

7 A discussion of the binocular field of vision, with its peculiar stereoscopic features, is omitted here, for although familiar to all, it is not as easy to describe, and cannot be represented by a single plane drawing.

8 J. Popper of Vienna.

9 Compare the note at the conclusion of my treatise, *Die Geschichte und die Wurzel des Satzes der Erhaltung der Arbeit,* Prague, Calve, 1872.

10 Similarly, *esprit de corps,* class bias, national pride, and even the narrowest minded local patriotism may have a high value, *for certain purposes.* But such attitudes will not be shared by the broad-minded inquirer, at least not in moments of research. All such egoistic views are adequate only for practical purposes. Of course, even the inquirer may succumb to habit. Trifling pedantries and nonsensical discussions, the cunning

appropriation of others' thoughts, with perfidious silence as to the sources, the metaphorical dysphagia suffered when recognition must be given, and the crooked illumination of others' performances when this is done, abundantly show that the scientist and scholar have also the battle of existence to fight, that the ways of science still lead to the mouth, and that the *pure* quest of knowledge in our present social conditions is still an ideal.

11 The ego is not a definite, unalterable, sharply-bounded unity. None of these attributes are important; for all vary even within the sphere of individual life; in fact their alteration is even sought after by the individual. *Continuity* alone is important. This view accords admirably with the position which Weismann has recently reached by biological investigations. ("Zur Frage der Unsterblichkeit der Einzelligen," *Biolog. Centralbl.,* Vol. IV., Nos. 21, 22; compare especially pages 654 and 655, where the scission of the individual into two *equal* halves is spoken of.) But this continuity is only a means of predisposing and of conserving what is contained in the ego. This content and not the ego is the principal thing. This content, however, is not confined to the individual. With the exception of some insignificant and valueless personal memories, it remains preserved in *others* even after the death of the individual. The *ego* is unsavable. It is partly the knowledge of this fact, partly the fear of it, that has given rise to the many extravagances of pessimism and optimism, and to numerous religious and philosophical absurdities. In the long run we shall not be able to close our eyes to this simple truth, which is the immediate outcome of psychological analysis. We shall then no longer place so high a value upon the ego, which even during the individual life greatly changes, and which, in sleep or during absorption in some idea, just in our very happiest moments, may be partially or wholly absent. We shall then be willing to renounce *individual* immortality, and not place more value upon the subsidiary elements than upon the principal ones. In this way we shall arrive at a freer and more enlightened view of life, which will preclude the disregard of other egos and the over-estimation of our own. [It will be seen from the above remarks that I consider that form of immortality alone as possessing reality and worth, which, with others, Dr. Paul Carus upholds, and which may be found in his discussions in *The Monist, The Open Court, Fundamental Problems,* etc. – Mach, 1895.]

12 The habit of treating the unanalysed ego-complex as an indiscerptible unity frequently assumes in science remarkable forms. First, the nervous system is separated from the body as the seat of the sensations. In the nervous system again, the brain is selected as the organ best fitted for this end, and finally, to save the supposed psychical unity, a *point* is sought in the brain as the seat of the soul. But such crude conceptions are hardly fit even to foreshadow the roughest outlines of what future research will do for the connexion of the physical and the psychical. The fact that the different organs of sensation and memory are physically *connected* with, and can be readily *excited* by, one another, is probably the foundation of the "psychical unity."

I once heard the question seriously discussed, "How the percept of a large tree could find room in the little head of a man?" Now, although this "problem" is no problem, yet it renders us vividly sensible of the absurdity that can be committed by thinking sensations spatially into the brain. When I speak of the sensations of *another* person, those sensations are, of course, not exhibited in my optical or physical space; they are mentally added, and I conceive them *causally,* not spatially, annexed to the brain observed or represented. When I speak of *my own* sensations, these sensations do not exist spatially in my head, but rather my "head" *shares* with them the same spatial field, as was explained above. (Compare the remarks on Fig. [12.1].)

[The extent to which the old notion of the soul still pervades modern physiological research, the purpose of which is precisely to overcome that ancient view, may be learned from Hauptmann's *Metaphysik in der Physiologie*, Dresden, 1893, with whose remarks I am in general accord. – Mach. 1895.]

13 I have always felt it as a stroke of special good fortune, that early in life, at about the age of fifteen, I lighted, in the library of my father, on a copy of Kant's *Prolegomena zu jeder künftigen Metaphysik.* The book made at the time a powerful and ineffaceable impression upon me, the like of which I never afterward experienced in any of my philosophical reading. Some two or three years later the superfluous rôle played by "the thing in itself" abruptly dawned upon me. On a bright summer day under the open heaven, the world with my ego suddenly appeared to me as *one* coherent mass of sensations, only more strongly coherent in the ego. Although the actual working out of this thought did not occur until a later period, yet this moment was decisive for my whole view. I had still to struggle long and hard before I was able to retain the new conception in my specialty. With the valuable parts of physical theories we necessarily absorb a good dose of false metaphysics, which it is very difficult to sift out from what deserves to be preserved, especially when those theories have become very familiar to us. At times, too, the traditional, instinctive views would arise with great power and place impediments in my way. Only by alternate studies in physics and in the physiology of the senses, and by historico-physical investigations (since about 1863), and after having endeavored in vain to settle the conflict by a physico-psycho logical monadology, have I attained to any considerable firmness in my views. I make no pretensions to the title of philosopher. I only seek to adopt in physics a point of view that need not be changed the moment our glance is carried over into the domain of another science; for, ultimately, all must form one whole. The molecular physics of to-day certainly does not meet this requirement. What I say I have probably not been the *first* to say. I also do not wish to offer this exposition of mine as a special achievement. It is rather my belief that every one will be led to a similar view, who makes a careful survey of any extensive body of knowledge. Avenarius, with whose works I recently became acquainted, approaches my point of view (*Philosophie als Denken der Welt nach dem Princip des kleinsten Kraftmasses,* 1876). Also Hering, in his paper on *Memory* (*Almanach der Wiener Akademie.* 1870, p. 258; English translation, O. C. Pub. Co., Chicago, 1895), and J. Popper in his beautiful book, *Das Recht zu leben und die Pflicht zu sterben* (Leipsic, 1878, p. 62), have advanced allied thoughts. Compare also my paper, *Ueber die ökonomische Natur der physikalischen Forschung* (*Almanach der Wiener Akademie,* 1882, p. 179, note; English translation in my *Popular Scientific Lectures,* Chicago, 1895). Finally let me also refer here to the introduction to W. Preyer's *Reine Empfindungslehre* and to Riehl's *Freiburger Antrittsrede,* p. 14. I should probably have much additional matter to cite as more or less allied to this line of thought, if my knowledge of the literature were more extensive.

14 [Molière's scourged philosopher (in *Le Mariage forcé*) does not say, It *seems* to me that I am pummelled, but, I *am* pummelled. – 1895.]

15 [A kindred view will be found in Avenarius (*Kritik der reinen Erfahrung,* and *Der menschliche Weltbegriff*). Avenarius has also undertaken the commendable task of explaining the development of philosophy on the basis of the facts furnished by the history of civilisation. For a further development of this view, which was evoked by a correspondence with Dr. Paul Carus, see the Appendix to this volume. – 1895.]

Editorial Headnote

13. Karl Pearson, *The Grammar of Science,* 1st edn (1892)

Karl Pearson (1857–1936) is a complex figure. He is now best known as a eugenicist and as one of the central figures in the development of mathematical statistics in the late nineteenth century. He also extensively developed a positivist philosophy of the physical sciences, in addition to, with W. F. R. Weldon (1860–1906), a mathematical approach to Charles Darwin's (1809–1882) nascent theory of evolution. This excerpt is from a work, *The Grammar of Science*, which Pearson wrote in 1892 to express his views in the philosophy of science. A second edition would add more work on evolution in 1900. Immediately influential, a young Albert Einstein would later arrange meetings of a reading group in Bern to discuss Pearson's book.

Pearson expresses a version of Mach's positivism that is perhaps even clearer than that of Mach. The scientific method consists, first, in the "careful and laborious classification of facts," then "the comparison of their relationships and sequences, and finally in the discovery by aid of the disciplined imagination of a brief statement or *formula*, which in a few words resumes the whole range of facts" (p. 227). These are *scientific laws*, which, in supporting the "economy of thought," will allow us "to make the fitting exertion on the receipt of a novel group of sense-impressions" (p. 227), that is, to be able to reason effectively on the basis of our future observations.

This leaves us with an interesting problem, however. If a scientific law is simply an aid to faster and more reliable mental calculation, then what is it that makes such a law special? We would like to think that Newton's law for planetary motion, for instance, possess a kind of "universal validity," or that it, as Pearson puts it, "will be accepted by every rational mind which has once understood its terms and clearly analyzed the facts which it resumes" (p. 228). A positivist concept of natural law seems too thin to be able to give us the tools that we need to recognize this special characteristic of laws of nature.

Pearson's central argumentative move is to qualify the "universal" in "universal validity." Just as the second law of thermodynamics would not be true for Maxwell's demon, or, he writes, as "our law of the tides could have no meaning for a blind worm on the shore, for whom the moon had no existence" (p. 232), "universal" validity seems to in fact mean only universal for humans like us with normally functioning perceptual apparatus. Since the motions of the planets had been the same before Newton formulated his law, it would be inaccurate, strictly speaking, to say that the planets were following Newton's law. That law depends upon the essentially human operations of classification and abstraction, and thus a part of its content "involves an association of natural facts or phenomena with mental conceptions, lying quite outside the particular field of those phenomena" (p. 232). That makes a natural law "a *résumé* in mental shorthand, which replaces for us a lengthy description of the sequences among our sense-impressions" (p. 233), with an ineliminable emphasis on the "mental." Echoing Comte, Pearson argues that scientific progress is equivalent to arriving at steadily more and more general instances of such scientific laws which are capable of describing – not "explaining," much less of "governing" – more and more features of the external world.

13

THE GRAMMAR OF SCIENCE

Karl Pearson

Source: Karl Pearson, *The Grammar of Science*, 1st edn (1892), pp. 92–104, 116–121.

CHAPTER III.

The Scientific Law.

§ 1. – *Résumé and Foreword.*

The discussions of our first two chapters have turned upon the nature of the method and material of modern science. The material of science corresponds, we have seen, to all the constructs and concepts of the mind. Certain of these constructs associated with immediate sense-impressions we project outwards and speak of as physical facts or phenomena; others, which are obtained by the mental processes of isolation and co-ordination from stored sense-impresses, we are accustomed to speak of as mental facts. In the case of both these classes of facts, the scientific method is the sole path by which we can attain to knowledge. The very word knowledge, indeed, only applies to the product of the scientific method in this field. Other methods, here or elsewhere, may lead to fantasy, as that of the poet or of the metaphysician, to belief or to superstition, but never to knowledge. As to the scientific method, we saw in our first chapter that it consists in the careful and often laborious classification[1] of facts, in the comparison of their relationships and sequences, and finally in the discovery by aid of the disciplined imagination of a brief statement or *formula*, which in a few words resumes the whole range of facts. Such a formula, we have, seen, is termed a *scientific law*. The object served by the discovery of such laws is the economy of thought; the suitable association of conceptions drawn from stored and correlated sense-impresses, permits the fitting exertion to follow with the minimum of thought upon the receipt of an immediate sense-impression. The knowledge of scientific law enables us to replace or supplement mechanical association, or instinct, by mental association, or thought. It is the *forethought*, by aid of which man in a far higher degree than other animals is able to make the fitting exertion on the receipt of a novel group of sense-impressions.

We are accustomed to speak of scientific law, or at any rate of one form of it termed "natural law," as something universally valid; we hold it to be as true for

DOI: 10.4324/9781003263814-18

all men as for its original propounder. Nay, there are not wanting those who assert that natural law has a validity quite independent of the human minds which formulate, demonstrate, or accept it. We can easily observe that there is really something *sui generis* about the validity of natural law. The philosopher, who propounds a new system, or the prophet who proclaims a new religion, may be absolutely convinced of the truth of his statement; but it is the result of experience from time immemorial that he cannot *demonstrate* that truth so that conviction is produced in the mind of every rational being. A philosophic or a religious formula – for example, the idealism of Berkeley, the scepticism of Hume, or the self-renunciation of the mediæval mystics, – however sure its teachers may be that it is capable of rational demonstration, really appeals to the individual temperament, and is accepted or rejected according to the emotional sympathies of the individual. On the other hand a formula, like that which Newton propounded for the motion of the planetary system, will be accepted by every rational mind which has once understood its terms and clearly analyzed the facts which it resumes.[2] This is sufficient to indicate that there must be some wide difference between philosophic and scientific systems, between theological and scientific formulae. I shall endeavour in this chapter to ascertain wherein this difference lies, to discover what is the meaning of the word law when used scientifically, and in what sense we can say that scientific law has universal validity.

§ 2. – *Of the Word Law and its Meanings.*

The term *law* probably recalls to the reader, in the first place, the rules of conduct proclaimed by the state and enforced under more or less heavy penalties against certain classes of its citizens. Austin, the most luminous English writer on jurisprudence,[3] who has devoted a very large portion of his well-known work to a discussion of the meaning of the word *law*, remarks: –

> "A law, in the most general and comprehensive acceptation in which the term, in its literal meaning, is employed, may be said to be a rule laid down for the guidance of an intelligent being by an intelligent being having power over him."

He further goes on to observe that where there is such a rule there is a command, and where there is a *command* a corresponding *duty*. From this standpoint Austin proceeds to discuss the various types of law, such as civil, moral, and divine law. It will be at once seen that with Austin's definition of law there is no place left for law in the scientific sense. He himself recognizes this, for he writes: –

> "Besides the various sorts of rules which are included in the literal acceptation of the term law, and those which are by a close and striking analogy, though improperly, termed laws, there are numerous applications of the term law, which rest upon a slender analogy and are

merely metaphorical or figurative. Such is the case when we talk of *laws* observed by the lower animals; of *laws* regulating the growth or decay of vegetables; of *laws* determining the movements of inanimate bodies or masses. For where *intelligence* is not, or where it is too bounded to take the name of *reason*, and therefore is too bounded to conceive the purpose of a law, there is not the *will* which law can work on, or which duty can incite or restrain. Yet through the misapplications of a *name,* flagrant as the metaphor is, has the field of jurisprudence and morals been deluged with muddy speculation" (p. 90).

Now Austin was absolutely in the right to emphasize the immense distinction between the use of the term *law* in science and its use in jurisprudence. There can be no doubt that the use of the same name for two totally different conceptions has led to a great deal of confusion. But on the one hand, if the flagrant misapplication of the scientific meaning of the word law to the fields of jurisprudence and morals has deluged them with "muddy speculation," there is equal certainty on the other hand that the misapplication of the legal and moral sense of the term has been equally disadvantageous to clear thinking in the field of science. Austin probably had in his mind when he wrote the above passage, works like Hegel's *Philosophy of Law*, in which we find the conception of the permanent and absolute character of scientific law applied to build up a system of absolute civil and moral law which somehow realizes itself in human institutions. To the mind which has once thoroughly grasped the principle of evolution in its special factor of natural selection, the civil and moral laws of any given society at a particular time must appear as ultimate results of the struggle for existence between that society and its neighbours. The civil and moral codes of a community at any time are those which are on the average best adapted to its current needs, and best calculated to preserve its stability. They are very plastic, and change in every age with the growth and variation of social conditions. What is lawful is what is not prohibited by the laws of a particular society at a particular time; what is moral is what tends to the welfare of a particular society at a particular time. We are all well acquainted with the continual change of civil law; in fact we keep up an important body, termed Parliament, whose chief function it is to modify and adapt our laws, so that they shall be best fitted at each period to assist the community in its struggle for existence. Of the changes in moral law we are, perhaps, less conscious, but they are none the less real. There are very few acts which have not been moral at some period in the growth of one or other society, and there are in fact many questions with regard to which our moral judgment is totally different from that of our grandfathers. It is the relativity, or variability with age and community, of civil and moral law, which led Austin, I think, to speak somewhat strongly of the speculation which confuses such law with law in the absolute sense of science. A law in the legal or moral sense holds only for individuals and individual communities, and is capable of modification or repeal. A law of science will be seen in the sequel to hold for all normal human beings so long as their perceptive and reasoning faculties remain without material modification. The confusion of these two ideas is productive of that "muddy speculation" which finds analogies between natural laws and those of the spiritual or moral world.

Now if we find that two quite distinct ideas unfortunately bear the same name we ought, in order to avoid confusion, to re-name one of them, or failing this we ought on all occasions to be quite sure in which of the two senses we are using the name. Accordingly, in my first chapter, in order to keep clear of the double sense of the word law, I endeavoured to replace it, when spoken of scientifically, by some such phrase as the "brief statement or formula which resumes the relationship between a group of facts." Indeed it would be well, were it possible, to take the term *formula*, as already used by theologians and mathematicians, and use it in place of scientific or natural law. But the latter term has taken such root in our language that it would be hard indeed to replace it now. Besides, if the word law is to be used in one sense only, we may ask why it is the scientist rather than the jurist who is to surrender his right to the word? The jurists say that historically they have the older claim to the word – that civil law existed long anterior to scientific law. This is perfectly true in a certain sense,[4] because the earliest attempts to codify laws for the conduct of men living in communities preceded any conscious recognition of scientific law. Now this leads us directly to a very important distinction, which, if it be neglected, is the source of much confusion. Does law exist before it receives expression and recognition? According to Austin, law in the juridical sense certainly does not, for such a law involves a "command," and a "corresponding duty" – that is, expression and recognition. What are we to say, then, with regard to scientific law – does it really exist before man has given expression to it? Has the word any meaning when unassociated with the mind of man? I hold that we must definitely answer "no" to both these questions, and I believe that the reader who has carefully followed my second chapter will see at once the grounds for this statement. A scientific law is related to the perceptions and conceptions formed by the perceptive and reasoning faculties in man; it is meaningless except in association with these; it is the *résumé* or *brief expression* of the relationships and sequences of certain groups of these perceptions and conceptions, and exists only when formulated by man.

§ 3. – *Natural Law relative to Man.*

Let us take that branch of scientific law which deals with the so-called "outside world" – natural law. We have seen that this outside world is a *construct.* It consists of objects constructed partly from immediate sense-impressions, and partly from stored impresses. For this reason the "outside world" is essentially conditioned by the perceptive and retentive faculties in man. Even the metaphysicians, who postulate "things-in-themselves," admit that sense-impressions in nowise *resemble* them, and that *man's* sense-impressions so far from being the entire product of "things-in-themselves," are probably but the smallest portion of their "capacity for producing" sense-impression. Hence to talk about natural law as existing in "things-in-themselves" and apart from man's mind is again to assert an unmeaning *x* among an unthinkable *y* and *z*. If nature for man is conditioned by his perceptive and retentive faculties, then natural law is conditioned by them also. It has no relation to something above and beyond man, but solely to the special products of his perceptive faculty. We have no right to infer its existence for things without a perceptive faculty, or even for perceptive faculties not closely akin to man's. I believe that a great deal of the obscurity involved in popular ideas about "Nature" would have been avoided had this been borne in mind.

A good instance of the *relativity* of natural law is to be found in the so-called *Second Law of Thermo-dynamics*. This law resumes a wide range of human experience, that is, of sequences observed in *our* sense-impressions, and embraces a great number of conclusions not only bearing on practical life, but upon that dissipation of energy which is even supposed to foreshadow the end of all life. The appreciation of the relativity of natural law is so important that the reader will, I trust, pardon me for citing the entire passage in which Clerk-Maxwell discusses this instance:[5] –

> "One of the best-established facts in thermo-dynamics is that it is impossible in a system enclosed in an envelope which permits neither change of volume nor passage of heat, and in which both the temperature and pressure are everywhere the same, to produce any inequality of temperature or of pressure without the expenditure of work. This is the second law of thermo-dynamics, and it is undoubtedly true so long as we can deal with bodies only in mass, and have no power of perceiving or handling the separate molecules of which they are made up. But if we conceive a being whose faculties are so sharpened that he can follow every molecule in its course, such a being, whose attributes are still as essentially finite as our own, would be able to do what is at present impossible to us. For we have seen that the molecules in a vessel of air at uniform temperature are moving with velocities by no means uniform, though the mean velocity of any great number of them, arbitrarily selected, is almost exactly uniform. Now let us suppose that such a vessel is divided into two portions, A and B, by a division in which there is a small hole, and that a being,[6] who can see the individual molecules, opens and closes this hole, so as to allow only the swifter molecules to pass from A to B, and only the slower ones to pass from B to A. He will thus, without expenditure of work, raise the temperature of B and lower that of A, in contradiction to the second law of thermo-dynamics."

To render this passage clear to the lay reader, we have only to add that in this kinetic theory the temperature of a gas depends upon the mean speed of its molecules. Now the second law of thermo-dynamics resumes with undoubted correctness a wide range of human experience, and is, to that extent, as much a law of nature as that of gravitation. But the kinetic theory of gases, whether it be hypothetical or not, enables us to conceive a demon having a perceptive faculty differing rather in degree than quality from our own, for whom the second law of thermodynamics would not necessarily be a law of nature. Such a conception enables us to grasp how relative what we term nature is to the faculty which perceives it. Scientific law does not, any more than sense-impression, lie in a universe outside and unconditioned by ourselves. Clerk-Maxwell's demon would perceive nature as something totally different from our nature, and to a less extent this is in great probability true for the animal world, and even for man in different stages of growth and civilization. The worlds of the child and of the savage differ widely from that of normal civilized man. One half of the perceptions which the latter links together in a law of nature may be wanting to the former. Our law of the tides could have no meaning for a blind worm on the shore, for whom the moon had no existence.[7] By the contents and the manner of perception the law of nature is essentially conditioned for each perceptive faculty. To speak, therefore, of the universal validity of a law of nature has only meaning in so far as we refer to a certain type of perceptive faculty, namely, that of a normal human being.

§ 4. – *Man as the Maker of Natural Law.*

The other problem with which we are concerned is the existence or non-existence of a scientific law before it has been postulated. Here the reader will feel inclined to remark: "Admitted that 'Nature' is conditioned by man's perceptive faculty, surely the sequences of man's perceptions follow the same law whether man has formulated that law in words or not? The law of gravitation ruled the motion of the planets ages before Newton was born." Yes and no, reader; the answer must depend on how we define our terms. The sequences involved in man's perception of the motion of the heavenly bodies were doubtless much the same to Ptolemy and Newton; to primitive man and to ourselves the motion of the sun is a common perception, but a sequence of sense-impressions is not in itself a law. That planets move, that a chick takes its origin from the egg, may be sequences of sense-impressions, they may be facts to be dealt with scientifically, but they are not laws in themselves, at least not in any useful interpretation of the word. The changes of the whole planetary system might be perceived, and even those perceptions translated into words with a fulness surpassing that of our most accurate modern observer, and yet neither the sequence of perceptions in itself nor the description involve the existence of any law. The sequence of perceptions has to be compared with other sequences, classification and generalization have to follow; conceptions and ideas, pure products of the mind, must be formed, before a description can be given of a range of sequences which, by its conciseness and comprehensiveness, is worthy of the name of scientific law.

Let it be noted that in this it is not only the process of reaching scientific law which is mental, but that the law itself when reached involves an association of natural facts or phenomena with mental conceptions, lying quite outside the particular field of those phenomena. Without the mental conceptions the law could not be, and it only comes into existence when these mental conceptions are first associated with the phenomena. The law of gravitation is not so much the discovery by Newton of a rule guiding the motion of the planets as his invention of a method of briefly describing the sequences of sense-impressions, which we term planetary motion. He did this in terms of a purely mental conception, namely, mutual acceleration.[8] Newton first brought the idea of mutual acceleration of a certain type into association with a certain range of phenomena, and was thus enabled to state a formula, which, by what we may term mental shorthand, resumes a vast number of observed sequences. The statement of this formula was not so much the discovery as the *creation* of the law of gravitation. A natural law is thus seen to be a *résumé* in mental shorthand, which replaces for us a lengthy description of the sequences among our sense-impressions. Law in the scientific sense is thus essentially a product of the human mind and has no meaning apart from man. It owes its existence to the creative power of his intellect. There is more meaning in the statement that man gives laws to Nature than in its converse that Nature gives laws to man.

[. . .]

§ 10. – *Progress in the Formulating of Natural Law.*

By the formation of conceptions, which may or may not have perceptual equivalents in the sphere of sense-impression, the scientist is able to classify and compare phenomena. From their classification he passes to formulae or scientific laws describing their sequences and relationships. The wider the range of phenomena embraced, and the simpler the statement of the law, the more nearly we consider that he has reached a "fundamental law of nature." The progress of science lies in the continual discovery of more and more comprehensive formulae, by aid of which we can classify the relationships and sequences of more and more extensive groups of phenomena. The earlier formulae are not necessarily wrong, they are merely replaced by others which in briefer language describe more facts.

We cannot do better than examine this process very briefly in a special case, namely, the motion of the planetary system. An easily observed part of this motion was the daily passage of the sun, its rising in the East and setting in the West. A primitive description of the motion consisted in the statement that the *same* sun which set in the West passed, hidden by northern mountains, along the surface of the *flat* earth and rose again in the East. The description was clearly very insufficient, but it was a first attempt at a scientific formula. An obvious improvement was soon made by limiting the surface of the earth and supposing the sun to go below the solid earth. The motion of the sun taken in conjunction with the motion of the stars led early astronomers to conclude that the earth was fixed in mid-space, and sun and stars were daily carried round it. The description thus improved was still far from complete; the sun was observed to vary its position with regard to the fixed stars. Gradually and laboriously facts were accumulated, and in time those early astronomers concluded that the sun went round yearly in the same circle, this circle itself being carried round with the starry heavens once in a day. This formula embraced a wider field of phenomena than the earlier ones, and probably was as exact a description as men's perceptions of earth and sun allowed when it was invented. Hipparchus improved it by placing the earth not exactly in the centre of the sun's circle, and thus more accurately described certain apparent irregularities in the sun's motion. A still more complete description was adopted by Ptolemy (A.D. 140) nearly three hundred years after Hipparchus, who, fixing the spherical earth, considered sun and moon to move in circles yearly round the earth, and the other planets in circles, whose centres again described circles round the earth. The whole of this system revolved daily round the earth with the stars. This, the famous Ptolemaic system, remained for many centuries the current formula, and even to this day the eccentrics of Hipparchus and epicycles of Ptolemy are not without service as elements of the more modern description. It would be wrong, I think, to say that the Ptolemaic system was an erroneous *explanation*, it was simply an insufficient attempt to *describe* in brief and accurate language a too limited range of phenomena. Then at the end of the Middle Ages came Copernicus who got rid of the cumbersome sphere carrying the fixed stars by simply considering the earth to rotate round its axis and of the epicycles, if not of the eccentrics, by treating the sun, not the earth, as the central point of the system. Here was an immense advance in brevity and accuracy of description; but still more facts remained to be included, more difficulties to be analyzed and overcome. This work was largely done by Keppler, who conceived the earth and planets to move in certain curves termed ellipses, of which the sun occupied a non-central point termed the focus. The formula of Keppler is one of the greatest achievements of the scientific method; it was the work of a disciplined imagination analyzing a laborious and minute classification of facts.[9] A more wide-embracing statement than that of Keppler was not only possible, however, but required; and this was provided by Newton in a single formula which embraces not only the motion of the planets, but that of their moons and of bodies at their surfaces. This formula is the well-known law of gravitation, but it is just as much a *description* of what takes place in planetary motion as Keppler's laws are a description – it is simply a briefer, more accurate, and more wide-embracing statement. The one can just as fitly as the other be termed a natural law.

The law of gravitation is a brief description of *how* every particle of matter in the universe is altering its motion with reference to every other particle. It does not tell us *why* particles thus move; it does not tell us *why* the earth describes a certain curve round the sun. It simply resumes, in a few brief words, the relationships observed between a vast range of phenomena. It economizes thought by stating in mental shorthand that routine of our perceptions which forms for us the universe of gravitating matter.

We have in the law of gravitation an excellent example of a scientific law. We see in its evolution the continual struggles of the human mind to reach a more and more comprehensive and exact formula, and at last Newton reaches one so simple and so wide-embracing that many have thought nothing further can be achieved in this direction. "Here," says Paul du Bois-Reymond, "is the limit to our possible knowledge." If the reader once grasps the characteristics of this law of Newton's he will understand the nature of all scientific law. Men study a range of facts – in the case of nature the material contents of their perceptive faculty – they classify and analyze, they discover relationships and sequences, and then they describe in the simplest possible terms the widest possible range of phenomena. How idle is it, then, to speak of the law of gravitation, or indeed of any scientific-law, as *ruling* nature. Such laws simply *describe*, they never *explain* the routine of our perceptions, the sense-impressions we project into an "outside world."

The scientific law, while thus the product of a rational analysis of facts, is always liable to be replaced by a wider generalization. Such replacement of one formula by another is indeed the regular course of scientific progress. The only final test we have of the truth of any law, of the sufficiency of its description, the only proof that our intellect has been keen enough to reach a formula extending to the whole range of facts it professes to resume, is the actual comparison of the results of the formula with the facts themselves – that is, historical observation or physical experiment. This test is all that marks the division between scientific hypothesis and scientific law, and the scientific law itself must, with every increase of our perceptive powers, return to the position of hypothesis and be anew put to the test of experience. Yet what philosophic system, what fantasy of the metaphysical mind in the region of the supersensuous has stood like Newton's formula of gravitation without the least change, the least variation in its statement, for more than two hundred years? Assuredly none; they have all shifted their ground with every advance of man's positive knowledge. They have not stood the test of experience; they are phantasms, not truth; for, as Sir John Herschel has said: –

> "The grand, and indeed only, character of truth is its capability of enduring the test of universal experience, and coming unchanged out of every possible form of fair dicussion."

§ 11. – *The Universality of Scientific Law.*

The universality, the absolute character, which we attribute to scientific law is really relative to the human mind. It is conditioned: –

1. By the perceptive faculty. The outside world, the world of phenomena, must be practically the same for all normal human beings.
2. By the reflective faculty. The processes of association and logical inference, and the inner world of stored impresses and conceptions must be practically the same for all normal human beings.

Now, when we classify a number of things together and give them the same name, we can only mean to signify that they closely resemble each other in structure and action. Hence when we speak of *human* beings we are referring to a class which in the normal civilized condition have perceptive and reflective faculties nearly akin. It is therefore not surprising that normal human beings perceive the same world of phenomena, and reflect upon it in much the same manner. The "universality" of natural law, the "absolute validity" of the scientific method, depends on the resemblance between the perceptive and reflective faculties of one human mind and those of a second. Human minds are, within limits, all receiving and sifting-machines of one type. They accept only particular classes of sense-impressions – being like automatic sweetmeat-boxes which if well constructed refuse to act for any coin but a penny – and having received their material they arrange and analyze it, provided they are in working order, in practically the same manner. If they do not arrange and analyze it in this manner we say, that the mind is disordered, the reason wanting, the person mad. The sense-impressions of a madman may be as much reality for him as our sense-impressions are for us, but his mind does not sift them in the normal human fashion, and for him, therefore, our laws of nature are without meaning.

Notes

1 The reader must be careful to recollect that *classification* is not identical with collection. It denotes the systematic association of kindred facts, the collection, not of all, but of relevant and crucial facts.

2 *One* system of planetary gravitation is accepted throughout the civilized world, but more than a dozen distinct theological systems and almost as many philosophical schools hardly suffice even for our own country.

3 *Lectures on Jurisprudence,* 4th ed. London, 1879.

4 For our final conclusions as to the historical right to the word, see p. 114.

5 *Theory of Heat*, 3rd ed. p. 308. Longmans, 1872.

6 This "being" has become known to fame as "Clerk-Maxwell's demon," but it must be noted that Clerk-Maxwell supposes the being's attributes "essentially finite as our own" – a peculiarity not usually associated with demons.

7 This point is well brought out by Prof. Lloyd Morgan in his *Animal Life and Intelligence*. After pointing out the widely different character of the sense organs in man and insects he continues: –

> "Remember their compound eyes with mosaic vision, coarser by far than our retinal vision, and their ocelli of problematical value, and the complete absence of muscular adjustments in either one or the other. Can we conceive that, with organs so different, anything like a similar perceptual world can be elaborated in their insect mind? I for one cannot. Admitting therefore that their perceptions may be fairly surmised to be analogous, that their world is the result of construction, I do not see how we can for one moment suppose that the perceptual world they construct can in any accurate sense be said to resemble ours" (pp. 298–9, 356–7, 361).

8 The reader will find mutual acceleration fully defined and discussed in Chapter VIII.

9 The elaborate observations of Tycho Brahé. Keppler not only stated the form of the planetary path, but the mode of its description in his famous three laws.

BIBLIOGRAPHY TO VOLUME I

A. Appadurai (ed.), *The Social Life of Things: Commodities in Cultural Perspective* (Cambridge: Cambridge University Press, 1986).

T. Arabatzis and D. Howard, 'Introduction: Integrated History and Philosophy of Science in Practice,' *Studies in History and Philosophy of Science* 50 (2014), pp. 1–3.

L. Badash, 'The Completeness of Nineteenth-Century Science,' *Isis* 63:1 (1972), pp. 48–58.

M. Black, 'The Identity of Indiscernibles,' *Mind*, 61:242 (1952), pp. 153–64.

P. J. Bowler, *Science for All: The Popularization of Science in Early Twentieth-Century Britain* (Chicago: University of Chicago Press, 2009).

D. Boyle, *Mary Shepherd: A Guide* (Oxford: Oxford University Press, 2023).

T. M. Brown, 'Medicine in the Shadow of the *Principia*,' *Journal of the History of Ideas* 48:4 (1987), pp. 629–48.

E. A. Burtt, *The Metaphysical Foundations of Modern Science*, 2nd edn (New York: Doubleday, 1932).

R. Callergård, *An Essay on Thomas Reid's Philosophy of Science* (Stockholm: Stockholm Universitet, 2006).

N. Cartwright, S. Psillos, and H. Chang, 'Theories of Scientific Method,' in M. J. Nye (ed.), *The Cambridge History of Science, Volume 5: The Modern Physical and Mathematical Sciences* (Cambridge: Cambridge University Press, 2002), pp. 19–35.

I. B. Cohen, 'The Case of the Missing Author: The Title Page of Newton's *Opticks* (1704), with Notes on the Title Page of Huygens's *Traité de la Lumière*,' in J. Z. Buchwald and I. B. Cohen (eds), *Isaac Newton's Natural Philosophy* (Cambridge, MA: MIT Press, 2001), pp. 15–45.

F. Copleston, *A History of Philosophy, Volume VIII: Modern Philosophy: Empiricism, Idealism, and Pragmatism in Britain and America* (New York: Doubleday, 1966).

M. Friedman (ed.), *Immanuel Kant: Metaphysical Foundations of Natural Science* (Cambridge: Cambridge University Press, 2004).

M. J. S. Hodge, 'Capitalist Contexts for Darwinian Theory: Land, Finance, Industry, and Empire,' *Journal of the History of Biology* 42:3 (2009), pp. 399–416.

N. Huggett, *Space from Zeno to Einstein* (Cambridge, MA: MIT Press, 1999).

M. Jammer, *Concepts of Force: A Study in the Foundations of Dynamics* (Cambridge, MA: Harvard University Press, 1957).

A. Janiak (ed.), *Isaac Newton: Philosophical Writings* (Cambridge: Cambridge University Press, 2004).

A. Kenny, *A New History of Western Philosophy, Volume IV: Philosophy in the Modern World* (Oxford: Clarendon Press, 2007).

J. Klein, 'Francis Bacon,' *Stanford Encyclopedia of Philosophy* (Fall 2020 Edition), at https://plato.stanford.edu/archives/fall2020/entries/francis-bacon/ [accessed 21 June 2024].

M. D. Laubichler and E. H. Davidson, 'Boveri's Long Experiment: Sea Urchin Merogones and the Establishment of the Role of Nuclear Chromosomes in Development,' *Developmental Biology* 314:1 (2008), pp. 1–11.

L. Laudan, 'Towards a Reassessment of Comte's 'Méthode Positive',' *Philosophy of Science*, 38:1 (1974), pp. 35–53.

T. Maudlin, 'Buckets of Water and Waves of Space: Why Spacetime Is Probably a Substance,' *Philosophy of Science*, 60:2 (1999), pp. 183–203.

R. McCormmach, 'Mr. Cavendish Weighs the World,' *Proceedings of the American Philosophical Society* 142:3 (1998), pp. 355–66.

T. McGrew, M. Alspector-Kelly, and F. Allhoff (eds), *Philosophy of Science: An Historical Anthology* (Malden, MA: Wiley-Blackwell, 2009).

R. F. McRae, 'Introduction,' in J. S. Mill, *A System of Logic, Books I–III* (Toronto: University of Toronto Press, 1974), pp. xxi–xlviii.

I. Newton, *Opticks* (New York: Dover, 1952).

R. Nichols and G. Yaffe, 'Thomas Reid,' in E. N. Zalta and U. Nodelman (eds), *The Stanford Encyclopedia of Philosophy* (Summer 2023 Edition), at https://plato.stanford.edu/archives/sum2023/entries/reid/ [accessed 16 July 2024].

C. S. Peirce, 'Methodology,' in J. M. Baldwin (ed.), Dictionary of Philosophy and Psychology (New York: Macmillan, 1901), at https://psychclassics.yorku.ca/Baldwin/Dictionary/ [accessed 13 January 2025].

C. H. Pence, 'John Herschel's Methodology in the Scientific Community,' in S. Case and L. M. Verburgt (eds), *The Cambridge Companion to John Herschel* (Cambridge: Cambridge University Press, 2024), pp. 257–79.

S. Ross, *Nineteenth-Century Attitudes: Men of Science* (Dordrecht: Kluwer, 1991).

R. Scholl, 'Inference to the Best Explanation in the Catch-22: How Much Autonomy for Mill's Method of Difference?,' *European Journal for Philosophy of Science* 5 (2015), pp. 89–110.

J. A. Secord, *Visions of Science: Books and Readers at the Dawn of the Victorian Age* (Chicago: University of Chicago Press, 2014).

L. J. Snyder, *The Philosophical Breakfast Club* (New York: Broadway, 2011).

S. Turner, 'Cause, Teleology, and Method,' in T. M. Porter and D. Ross (eds), *The Cambridge History of Science, Volume 7: The Modern Social Sciences* (Cambridge: Cambridge University Press, 2003), pp. 57–70.

L. M. Verburgt, 'Scientific Method, Induction, and Probability: The Whewell-de Morgan Debate on Baconianism, 1830s–1850s,' *HOPOS* 14:1 (2024), pp. 134–163.

L. M. Verburgt, 'Herschel's Philosophy of Science: The Preliminary Discourse on the Study of Natural Philosophy and Beyond,' in S. Case and L. M. Verburgt (eds), *The Cambridge Companion to John Herschel* (Cambridge: Cambridge University Press, 2024), pp. 99–126.

D. B. Wilson, 'Herschel and Whewell's Version of Newtonianism,' *Journal of the History of Ideas* 35:1 (1974), pp. 79–97.

J. Wilson, "On Mary Shepherd's *Essay Upon the Relation of Cause and Effect*," in E. Schliesser (ed.), *Neglected Classics of Philosophy, Volume 2* (Oxford: Oxford University Press, 2022), pp. 141–171.

INDEX

FSC
www.fsc.org
MIX
Papier | Fördert gute Waldnutzung
FSC® C083411

Kolibri 360 GmbH
Ferdinand-Jühlke-Straße 7
99095 Erfurt, Deutschland
produktsicherheit@kolibri360.de